Planning and Operation of Plug-In Electric Vehicles

Mehdi Rahmani-Andebili

Planning and Operation of Plug-In Electric Vehicles

Technical, Geographical, and Social Aspects

Springer

Mehdi Rahmani-Andebili
Department of Physics and Astronomy
University of Alabama in Huntsville
Huntsville, AL, USA

ISBN 978-3-030-18021-8 ISBN 978-3-030-18022-5 (eBook)
https://doi.org/10.1007/978-3-030-18022-5

This Springer imprint is published by the registered company Springer Nature Switzerland AG
The registered company address is: Gewerbestrasse 11, 6330 Cham, Switzerland

Preface

This book studies the operation and planning problems of plug-in electric vehicles (PEV) while considering the social and geographical aspects of problems and their technical characteristics.

A significant portion of energy consumption and greenhouse gas emissions are attributed to the transportation sector. Transportation electrification is seen as one of the solutions to the above-mentioned challenges to mitigate the geopolitical concerns about the availability of fossil fuels and global warming. Due to the economic and environmental benefits of electric vehicles and PEVs, they are gaining more popularity in the auto-market, since they can be charged by the renewable energy sources as the clean and free sources of energy.

Governments worldwide are implementing financial incentives to accelerate the transition from the internal combustion engine vehicles to the electric ones to achieve their own energy security and climate change mitigation goals. On the other hand, as the battery prices decrease, the unsubsidized electric vehicles are made as affordable as their gasoline counterparts.

A number of research organizations predict that the electric vehicles' (including PEVs) sales will surpass the gasoline and diesel vehicles' sales in the near future. Therefore, the PEVs' drivers, as the new electricity customers, will consume a considerable portion of electricity in the future. Although a single PEV does not have a remarkable effect on the electric power system, the uncontrolled charging/discharging pattern of a large number of PEVs has a potential to negatively affect the power system's performance and put it at a risk. In other words, the uncoordinated charging/discharging of PEVs may cause several serious issues for different parts of power system such as causing an overloading in the generation system, a congestion in the transmission and distribution lines, and spikes in the electricity market prices. Therefore, the fleet management of PEVs in a power system is unavoidable and even desirable.

The PEVs are parked at the parking lots in the idle status for most of the day. Thus, a large size of electric power and capacity will be available. In this regard, the power system operators and PEV aggregators can play a key role to mitigate the above-mentioned challenges and issues, since they can motivate the PEV drivers, by offering a variety of incentives to them, to let them manage their vehicles' charging time considering the requested expectation of drivers to have a desirable charge level at the departure time. By implementing this plan, the operators and aggregators can take part in different activities in the power markets and provide the benefits for themselves, for the electric power system, and even for the PEV owners.

In order to model the social characteristics of drivers from the sociology and economic points of view, the drivers are stratified in three different social classes based on their income level including low-income, moderate-income, and high-income. This type of classification is necessary to be considered in the modelling of drivers' interaction with the power system operators and PEV aggregators, since the drivers, belonging to different social classes, dissimilarly react with respect to the value of incentive. In other words, this social classification helps determine the responsiveness degree of drivers about the cooperation request received from an agent in the smart grid. Generally, the low-income and high-income drivers are more and less likely to be interested in the proposed incentive, respectively. This assumption is applied to model the social aspect of problems studied in the book.

Furthermore, in order to model the geographical features of problems, the driving routes and the hourly distance of PEVs from the charging stations or parking lots are modelled. This kind of modelling is obligatory, since the responsiveness probability of drivers (even belonging to the same social class) with respect to the value of incentive will not be equal if they have different distances from a charging station.

The technical specifications of PEVs that have been considered in the book chapters include the hourly state of charge of a PEV, the depth of discharge limit defined for a PEV, the various PEV types with different rated powers and capacities, the efficiency of a PEV's battery, the capability of a PEV for the vehicle-to-grid and grid-to-vehicle services provision, the PEV's battery degradation, the PEV parking lot power and capacity, and the yearly maintenance of PEV parking lot.

Two types of problem including the planning and operation problems are studied in the book. The first five chapters are concerned with the operation problems; however, the sixth chapter covers the planning problem, and a data analysis technique is presented in the seventh chapter. In this book, the operation problems are studied in the level of both electrical generation and distribution systems, while the planning problem is concerned with the electrical distribution system.

Those problems that deal with the electrical generation system (Chaps. 1, 2, and 3) consider the technical and social aspects of problem and ignore the geographical aspect, since the location of vehicle-to-grid and grid-to-vehicle services provision does not matter for a generation company. However, the problems related to the electrical distribution system (Chaps. 4, 5, and 6) take into account the geographical characteristics of area, in addition to the technical specifications of PEVs and the social aspect of drivers.

The problems that deal with the generation system are studied from a generation company's point of view and investigate the effects of optimal fleet management of PEVs on the real power markets' demands, unit commitment problem, and spinning reserve capacity provision problem. Moreover, the problems concerning with the electrical distribution system are investigated from a distribution company's viewpoint and cover the robust operation of distribution system, optimal operation of a PEV parking lot in the energy market, and optimal placement and sizing of PEV parking lots by optimal fleet management of PEVs.

In the preparation of book chapters, it has been tried to apply a variety of optimization techniques including genetic algorithm, improved genetic algorithm, mixed-integer linear programming, and simulated annealing. In addition, Chaps. 6 and 7, which take into account the geographical aspect of problem, apply the real driving routes and the real geography of area (i.e., latitude and longitude data) of San Francisco, CA, USA. Moreover, Chaps. 4 and 5, which include the geographical aspect of problem, consider the real geography of areas (i.e., latitude and longitude data) of Tehran, Iran (around Sharif University of Technology), and Marina City Vertical Parking Lot, Chicago, IL 60654, USA, respectively. However, the driving routes have been designed by a computer considering the related minimum and maximum traffic speed limits.

In each chapter, a variety of sensitivity analyses have been performed to study the effects of input parameters on the output parameters. The input parameters include the social class of drivers (low-income, moderate-income, and high-income), the PEV penetration level (low, moderate, and high), the PEV type (Tesla Model S, Citroën C-Zero, Volkswagen e-Up, and Renault Kangoo Z.E.), the feeder's failure rate model, and the voltage-dependent load model (a combination of residential, commercial, and industrial loads). In addition, the output parameters include the maximum load factor of system, the optimal generation scheduling and commitment of generation units, the optimal status of feeders' switches (open and close), the optimal mode of energy storage systems (charging, discharging, and idle), the maximum profit of parking lot, the optimal location and size of parking lots, the optimal value of incentive, the minimum total cost of planning problem, the optimal value of energy loss and reliability indices of system, the optimal voltage profile of buses, and the optimal power flow of feeder's branches.

The students and scholars are welcome to send their comments to the author of the book if they find any misprint or any other possible mistakes in the book chapters. Their name and contribution will be mentioned in the next version of book.

Huntsville, AL, USA Mehdi Rahmani-Andebili

Contents

Chapter 1
Studying the Effects of Plug-In Electric Vehicles on the Real Power Markets Demand Considering the Technical and Social Aspects

1.1 Introduction

Almost one-third of total energy consumption and greenhouse gas emissions in the world are related to the transportation sector [1]. Figure 1.1 shows the air pollution caused by the internal combustion-based vehicles in Donora, PA, USA, on Sep. 19, 2017 [2].

Due to the economic and environmental benefits of plug-in electric vehicles (PEVs), PEVs are gaining more popularity in the auto-market [3, 4], since they can be charged by the renewable energy sources (RES) as the clean and free sources of energy. As demand and interest in PEVs increase, more and more manufacturers are starting to produce and sell PEVs. Figure 1.2 illustrates a Tesla Supercharger station in Atlanta, GA, USA [5].

Figure 1.3 shows the map of fuel cost savings per year for a vehicle in each state of the USA, when driving on electricity instead of gas [6]. As can be seen, there is a considerable potential to minimize the fuel consumption expenses. Reference [7] states that Washington DC residents will get a tax credit of $19,000 per vehicle if they convert their vehicles' fuel consumption to the alternative ones including electricity.

PEVs can act as the energy storages and shift part of the demand from the peak period to the off-peak period. Therefore, PEV fleet can improve the market load factor by leveling the demand profile if an optimal fleet management (FM) is applied by the system operator.

FM of PEVs can be considered as one of the demand side management (DSM) measures. DSM is the first precedence in the energy policy decisions due to its benefits from economic and environmental viewpoints [8]. DSM provides the short-term responses to the electricity market conditions to reduce total cost of energy supply, increase reserve margin, mitigate price volatility, and achieve environmental goals by deferring the commitment of polluted generation units leading to increased energy efficiency and reduced greenhouse gas emissions [9]. According

M. Rahmani-Andebili, *Planning and Operation of Plug-In Electric Vehicles*,
https://doi.org/10.1007/978-3-030-18022-5_1

Fig. 1.1 Donora, PA, USA, on Sep. 19, 2017 [2]

Fig. 1.2 A Tesla Supercharger station in Atlanta, GA, USA [5]

to the Federal Energy Regulatory Commission (FERC), demand response (DR) is defined as:

- The changes in electricity usage by the end-use customers from their normal consumption pattern in response to the changes in the price of electricity over time, or to the incentive payments designed to induce lower electricity use at the times of high wholesale market prices or when the system reliability is jeopardized [10].

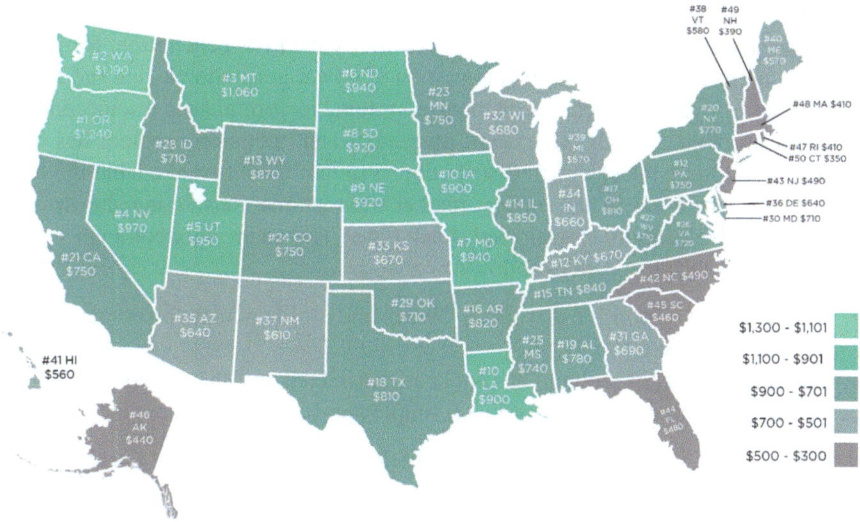

Fig. 1.3 Map of fuel cost savings per vehicle in the USA ($/year) when driving on electricity instead of gas [6]

Several studies have investigated the presence of PEVs in the electric power systems [11–23]. In [11], the planning and operation problems of parking lots have been solved by the distribution company (DISCO) and generation company (GENCO) considering the behavioral model of drivers and several economic and technical factors. In [12], a stochastic model predictive (MPC) control approach for the charging management of PEVs and the system reconfiguration has been presented considering the driving patterns and the drivers' behavioral models.

In [13], a PEV charging management framework for the utilization of photovoltaic (PV) system has been presented considering the possibility of information exchange between the home and grid. Reference [14] has presented a charging strategy that offers the multiple charging options to the PEVs including AC level II charging, DC fast charging, and battery swapping facilities. In [15], the stochastic presence of an aggregator, owning some parking lots, in the energy market transactions has been studied. In this study, the drivers' behavior, the hourly changing capacity of parking lots, the degradation of PEVs' batteries, and some economic factors such as inflation and interest rates have been considered.

In [16], a dynamic pricing and energy management problem have been solved for the PEVs' charging service providers considering the renewables and energy storage systems. In [17], a framework to determine the appropriate level of PEV penetration for the given electrical distribution system has been presented. In [18], the optimal parking lot allocation problem has been solved to minimize the total cost of local DISCO considering the vehicles traffic and the driving patterns of PEVs' owners. In [19], an algorithm to aggregate the flexible loads with the PEVs demand has been presented.

Reference [20] has presented a locational marginal pricing approach to mitigate the congestion of electrical distribution network by the PEV penetration. Reference [21] has performed a study on the integration of power generation assets and PEVs' drivers, as a cyber-physical-social power system, to formulate a socioeconomic dispatch problem. In [22], SmartPark placement and operation problem has been studied to improve the electrical distribution system reliability and enable the local DISCO to participate in the energy market transactions considering the drivers' behavior and electricity consumers' load type. The main shortcomings of the abovementioned studies are as follows:

- The impact of responsive drivers on the demand profile of real power markets has not been studied.
- The social aspect of problem has not been considered, and the drivers' behavioral model has been disregarded or just a simple approach has been applied.

Additionally, in [23], just the impacts of DR programs on some of the real power markets have been studied. The contributions of this chapter include:

- Modelling and studying the impact of PEVs on the demand profile of some of the well-known real power markets
- Implementing the optimal FM of PEVs considering different PEV penetration levels
- Categorizing the drivers in three different social classes based on their income level
- Modelling the behavior of each social class of drivers based on the drivers' reaction with respect to the value of incentive suggested by the independent system operator (ISO) to the drivers to transfer their charging demand from the peak period to the off-peak one
- Performing the sensitivity analysis with respect to the value of incentive and social class of drivers
- Studying the effects of unrealistic modelling of drivers' social class in each power market.

The remainder of chapter is organized as follows. In Sect. 1.2, the proposed approach that includes the social and technical aspects of problem is presented and described. In Sect. 1.3, the problem is simulated, and the results are analyzed and interpreted. Section 1.4 concludes the chapter.

1.2 Modelling the Social and Technical Aspects of Problem

1.2.1 Social Stratification of Drivers

In this study, the drivers are stratified in three different social classes based on their income level, namely, low-income (LI), moderate-income (MI), and high-income (HI). From sociology and economic points of view, this type of classification is necessary, since the drivers concerned with different social classes will dissimilarly react with respect to the value of incentive suggested by a power market operator. In other words, the low-income drivers are more likely to be interested in the proposed incentive compared to the high-income drivers.

The reason that the drivers are socially classified in this chapter is to model their willingness or reluctance level while being called by the power market operator. Table 1.1 mathematically expresses the responsiveness probability of low-income, moderate-income, and high-income drivers with respect to the value of incentive (discount on charging fee (β) in percent). Moreover, Fig. 1.4 graphically shows the responsiveness probability of a driver in each social class for the range of $0\% \leq \beta \leq 100\%$. The models present the responsiveness probability of a driver (with any social class) to transfer his/her charging demand from the peak period to the off-peak period. As can be seen in Table 1.1 and Fig. 1.4, there are some reasonable criteria behind these modelling presented as follows:

- The probability that a driver (with any social class) will transfer his/her charging demand from the peak period to the off-peak period is 0% if there is no incentive ($\beta = 0\%$).

Table 1.1 Responsiveness probability of a low-income, moderate-income, and high-income driver with respect to the value of incentive (discount on charging fee (β))

Social class	Driver's responsiveness probability (%)
Low-income (LI) drivers	$\xi_{e,LI,\beta,t} = 100 \times \left(\frac{\beta}{100}\right)^{0.3}$ (1.1)
Moderate-income (MI) drivers	$\xi_{e,MI,\beta,t} = 100 \times \left(\frac{\beta}{100}\right)$ (1.2)
High-income (HI) drivers	$\xi_{e,HI,\beta,t} = 100 \times exp\left(10 \times \left(\frac{\beta}{100} - 1\right)\right)$ (1.3)

Fig. 1.4 The graphical illustration for the responsiveness probability of a low-income, moderate-income, and high-income driver with respect to the value of incentive (β) for the range of $0\ \% \leq \beta \leq 100\%$

- The probability that a driver (with any social class) will transfer his/her charging demand from the peak period to the off-peak period is 100% if the PEV is charged in free ($\beta = 100\%$).

Herein, t and e are the indices of time and PEV, respectively.

The expected number of responsive drivers ($\tilde{N}^{PEVs}$) that will transfer their charging demand from the peak period to the off-peak period is determined using Eq. (1.4). Herein, SC and $N^{PEVs,\ TOTAL}$ are the index of social class and the total number of PEVs in the area, respectively.

$$\tilde{N}^{PEVs}_{SC,\beta,t} = \sum_{e=1}^{N^{PEVs,TOTAL}} \frac{\xi_{e,SC,\beta,t}}{100}, \quad \forall SC \in \{LI, MI, HI\}, \qquad (1.4)$$
$$\forall \beta \in \{0, 10, \ldots, 100\}, \quad \forall t \in \{1, \ldots, 24\}$$

1.2.2 Technical Aspect

Figure 1.5 illustrates the hourly demand profile (in per unit (p.u.)) of some of the real power markets including Electric Reliability Council of Texas (ERCOT), New York Independent System Operator (NYISO), Pennsylvania-Jersey-Maryland (PJM), and Independent System Operator New England (ISO-NE) [24]. In this chapter, to maintain the comprehensiveness of modellings and simulations, all the values are presented in p.u. and all the calculations are done in p.u. system. Herein, the peak and off-peak periods are determined using (1.5). As can be seen, the demand above 0.9 p.u. is considered as the peak period in any power market. Based on this definition, the time interval of each period in each power market is determined and presented in Table 1.2.

Moreover, the load factor of a power market is defined as Eq. (1.6). The primary load factor of each power market is presented in Table 1.2. As can be seen, ERCOT and ISO-NE power markets have the largest and the smallest load factors, respectively, among the power markets under study in this chapter.

$$Period = \begin{cases} Off\text{-}peak & D^{MARKET} < 0.9 \\ Peak & D^{MARKET} \geq 0.9 \end{cases}, \quad \forall MARKET \in \{ERCOT, NYISO, PJM, ISO\text{-}NE\} \qquad (1.5)$$

$$LF = 100 \times \frac{1}{Max\{D_t^{MARKET} | t \in 1, \ldots, 24\}} \times \frac{1}{24} \sum_{t=1}^{24} D_t^{MARKET} \qquad (1.6)$$

In [12], to model the variability and uncertainty of state of charge (SOC) of PEV fleet, a stochastic MPC approach has been applied. In this reference, the hourly SOC of PEV fleet has been modelled by the mean (μ) and standard deviation (σ) of hourly

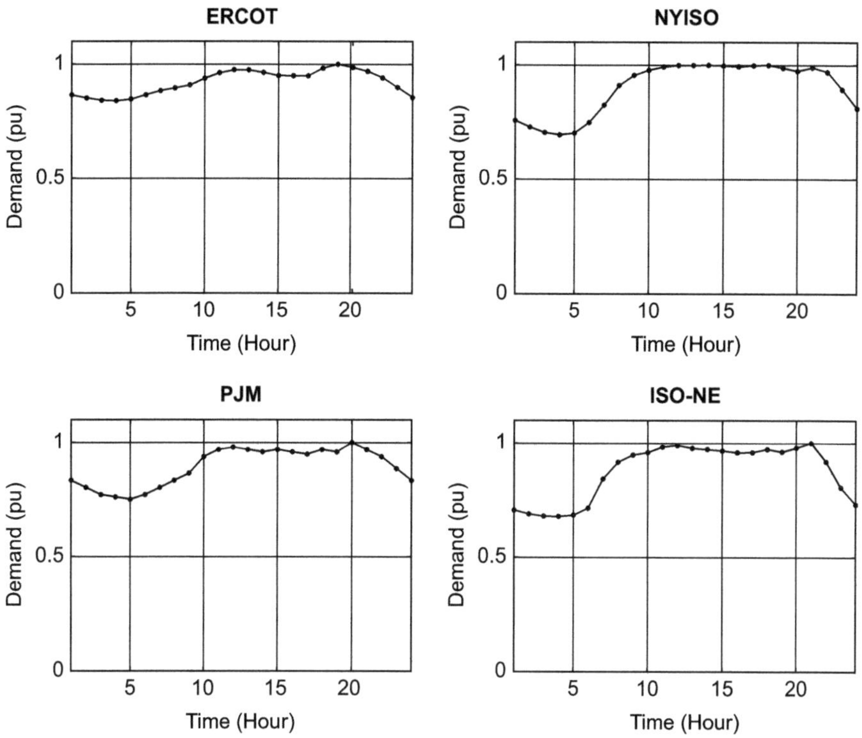

Fig. 1.5 Hourly demand profile (p.u.) of (**a**) ERCOT, (**b**) NYISO, (**c**) PJM, and (**d**) ISO-NE power markets [24]

Table 1.2 Initial load factor and interval of off-peak and peak periods in each power market

Power market	Off-peak period	Peak period	Initial load factor (%)
ERCOT	Hours: 1–8 and 23–24	Hours: 9–22	92.11
NYISO	Hours: 1–7 and 23–24	Hours: 8–22	90.06
PJM	Hours: 1–9 and 23–24	Hours: 10–22	89.28
ISO-NE	Hours: 1–7 and 23–24	Hours: 8–22	87.49

Gaussian distribution function. Figure 1.6a shows the Gaussian distribution functions concerned with the hourly SOC of PEV fleet [12]. The Gaussian distribution function related to the SOC of PEV fleet at 21st hour of the day is illustrated in Fig. 1.6b [12]. Herein, 15.9%, 34.1%, 34.1%, and 15.9% are the occurrence probabilities of SOC of PEV fleet corresponding to the values of $\mu - 2\sigma$, $\mu - \sigma$, $\mu + \sigma$, and $\mu + 2\sigma$, respectively. In Fig. 1.6, some logics can be identified presented as follow:

- The SOC of PEV fleet has a descending trend throughout the day.
- The SOC of PEV fleet has smaller standard deviations during the intervals of 1–7 and 22–24.
- The SOC of PEVs has larger standard deviations during the period of 8–21.

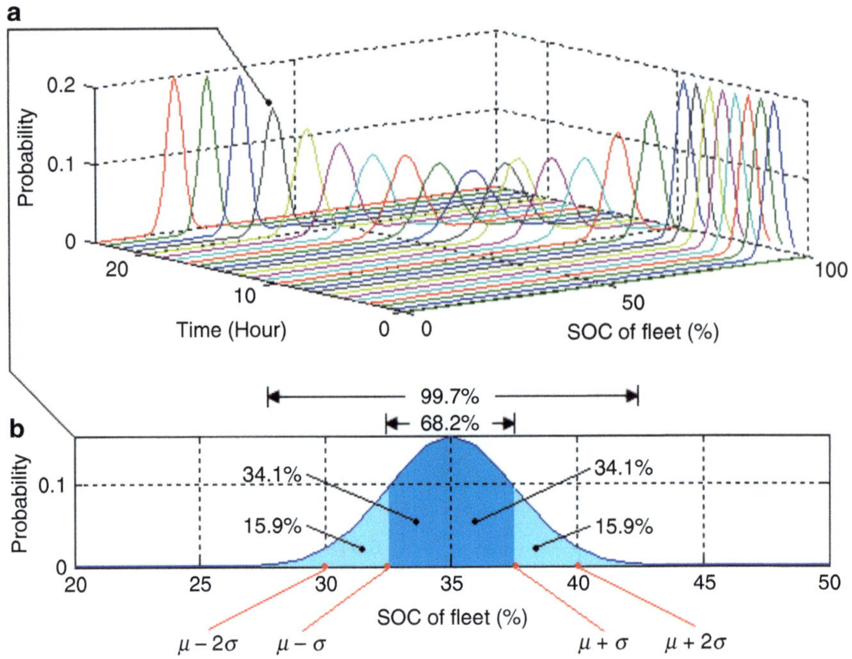

Fig. 1.6 (**a**) Gaussian distribution functions related to the hourly SOC of PEV fleet. (**b**) Gaussian distribution function related to the SOC of PEV fleet at 21st hour of the day [12]

In this chapter, the mean value of each hourly Gaussian distribution function is applied as the hourly SOC of PEV fleet. Equation (1.7), in MW, presents the charging demand of responsive PEV fleet ($\tilde{D}_{SC,\beta,t}^{PEVs}$) with social class of SC and incentive of β (discount of charging fee) at time t. Herein, SOC^{PEVs}, P^{PEV}, and $\tilde{N}^{PEVs}$ indicate the SOC of PEV fleet (percent), rated power of battery of a PEV (kW), and expected number of responsive drivers that will transfer their charging demand from the peak period to the off-peak period, respectively. In Eq. (1.8), $D_t^{PEVs,TOTAL}$, in MW, gives the total demand of PEVs (including responsive and nonresponsive PEVs), where $N^{PEVs,\ TOTAL}$ is the total number of PEV fleet in the area. Moreover, D_t^{PEVs} in Eq. (1.9) presents the hourly charging demand of nonresponsive PEV fleet (MW).

$$\tilde{D}_{SC,\beta,t}^{PEVs} = \left(1 - \frac{SOC_t^{PEVs}}{100}\right) \times \frac{P^{PEV}}{1000} \times \tilde{N}_{SC,\beta,t}^{PEVs} \tag{1.7}$$

$$D_t^{PEVs,TOTAL} = \left(1 - \frac{SOC_t^{PEVs}}{100}\right) \times \frac{P^{PEV}}{1000} \times N^{PEVs,TOTAL} \tag{1.8}$$

$$D_t^{PEVs} = D_t^{PEVs,TOTAL} - \tilde{D}_{SC,\beta,t}^{PEVs} \tag{1.9}$$

Equations (1.10, 1.11, and 1.12) present the p.u. value of Eqs. (1.7, 1.8, and 1.9), respectively. Total demand of other loads ($D_t^{LOADs}(p.u.)$) in p.u. is determined using (1.13), where $D_t^{MARKET}(p.u.)$ is the total demand of power market (p.u.) and φ^{PEVs} is the PEV penetration percentage in the related power market. The value of φ^{PEVs} for low, moderate, and high penetration levels are considered about 10%, 20%, and 30%, respectively.

Figure 1.7 illustrates the hourly total demands of PEV fleet before FM of the PEVs considering low, moderate, and high PEV penetration levels. The hourly demands have been calculated using the hourly SOC of PEV fleet given in Fig. 1.6 and assuming Nissan Leaf as the type of PEVs with the power and capacity of 24 kW and 24 kWh, respectively [25].

$$\tilde{D}_{SC,\beta,t}^{PEVs}(p.u.) = \frac{\tilde{D}_{SC,\beta,t}^{PEVs}}{Max\{D_t^{PEVs,TOTAL}|t \in 1,\ldots,24\}} \tag{1.10}$$

$$D_t^{PEVs,TOTAL}(p.u.) = \frac{D_t^{PEVs,TOTAL}}{Max\{D_t^{PEVs,TOTAL}|t \in 1,\ldots,24\}} \tag{1.11}$$

$$D_t^{PEVs}(p.u.) = \frac{D_t^{PEVs}}{Max\{D_t^{PEVs,TOTAL}|t \in 1,\ldots,24\}} \tag{1.12}$$

$$D_t^{LOADs}(p.u.) = D_t^{MARKET}(p.u.) - \frac{\varphi^{PEVs}}{100} \times D_t^{PEVs,TOTAL}(p.u.) \tag{1.13}$$

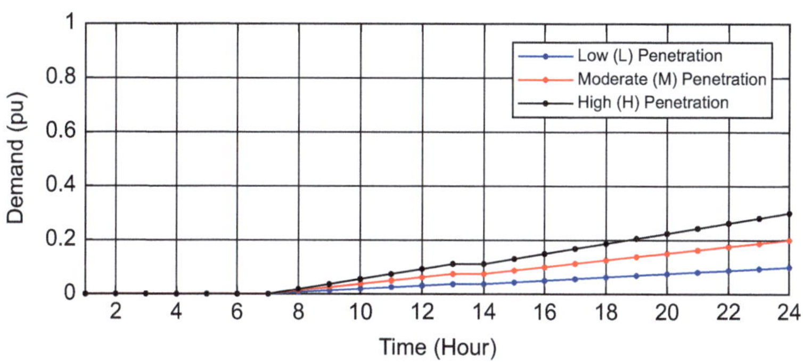

Fig. 1.7 Hourly total demand of PEV fleet before FM of PEV fleet considering low (L), moderate (M), and high (H) PEV penetration levels

1.3 Simulation Results

1.3.1 Base Case

In this section, the optimal value of incentive and the maximum value of load factor in each power market are investigated for various PEV penetration levels (low, moderate, and high), and the results are compared with the consequences of problem simulation without FM. In this part, the social class of drivers is moderate-income in each power market.

Tables 1.3, 1.4, 1.5, and 1.6 present the optimal value of incentive and the maximum value of market load factor before and after optimal FM of moderate-income drivers considering low, moderate, and high PEV penetration levels in ERCOT, NYISO, PJM, and ISO-NE power markets. Figures 1.8, 1.9, 1.10, and 1.11 show the hourly PEVs demand, loads demand, and total demand related to each power market before and after optimal FM of moderate-income drivers considering low, moderate, and high PEV penetration levels and their corresponding optimal incentive.

As can be seen in Tables 1.3, 1.4, 1.5, and 1.6 and Figs. 1.8, 1.9, 1.10, and 1.11, after optimal implementation of FM, the load factor of each power market increases

Table 1.3 Optimal value of incentive and maximum value of market load factor before and after optimal FM of moderate-income drivers for different PEV penetration levels in ERCOT power market

Market	Outputs	FM	Low penetration	Moderate penetration	High penetration
ERCOT	Optimal discount (%)	Before FM	0		
		After FM	70	40	20
	Optimal load factor (%)	Before FM	92.11		
		After FM	96.57	96.51	96.07

Table 1.4 Optimal value of incentive and maximum value of market load factor before and after optimal FM of moderate-income drivers for different PEV penetration levels in NYISO power market

Market	Outputs	FM	Low penetration	Moderate penetration	High penetration
NYISO	Optimal discount (%)	Before FM	0		
		After FM	100	50	30
	Optimal load factor (%)	Before FM	90.06		
		After FM	92.78	92.78	92.75

Table 1.5 Optimal value of incentive and maximum value of market load factor before and after optimal FM of moderate-income drivers for different PEV penetration levels in PJM power market

Market	Outputs	FM	Low penetration	Moderate penetration	High penetration
PJM	Optimal discount (%)	Before FM	0		
		After FM	100	50	30
	Optimal load factor (%)	Before FM	89.28		
		After FM	94.18	94.18	93.87

Table 1.6 Optimal value of incentive and maximum value of market load factor before and after optimal FM of moderate-income drivers for different PEV penetration levels in ISO-NE power market

Market	Outputs	FM	Low penetration	Moderate penetration	High penetration
ISO-NE	Optimal discount (%)	Before FM	0		
		After FM	100	70	40
	Optimal load factor (%)	Before FM	87.49		
		After FM	91.28	92.03	91.79

for any PEV penetration level. Moreover, almost different values of optimal incentive are determined for them. In addition, in each market, the case with low/high PEV penetration level needs more/less incentive to motivate the responsive drivers to transfer part of their charging demand from the peak period to the off-peak one and make the market demand profile flatter.

1.3.2 Sensitivity Analysis with Respect to Value of Incentive

In this part of chapter, a sensitivity analysis is done for the load factor of each power market with respect to the value of incentive for the range of 0–100%. In this section, like Sect. 1.3.1, the social class of drivers is moderate-income. As can be seen in Fig. 1.12, the optimal value of incentive is different for each market and even for a market with different PEV penetration levels. In NYISO, PJM, and ISO-NE, the responsive drivers' vehicles need to be charged in free (100% discount on charging fee) to have the maximum load factor for the low PEV penetration.

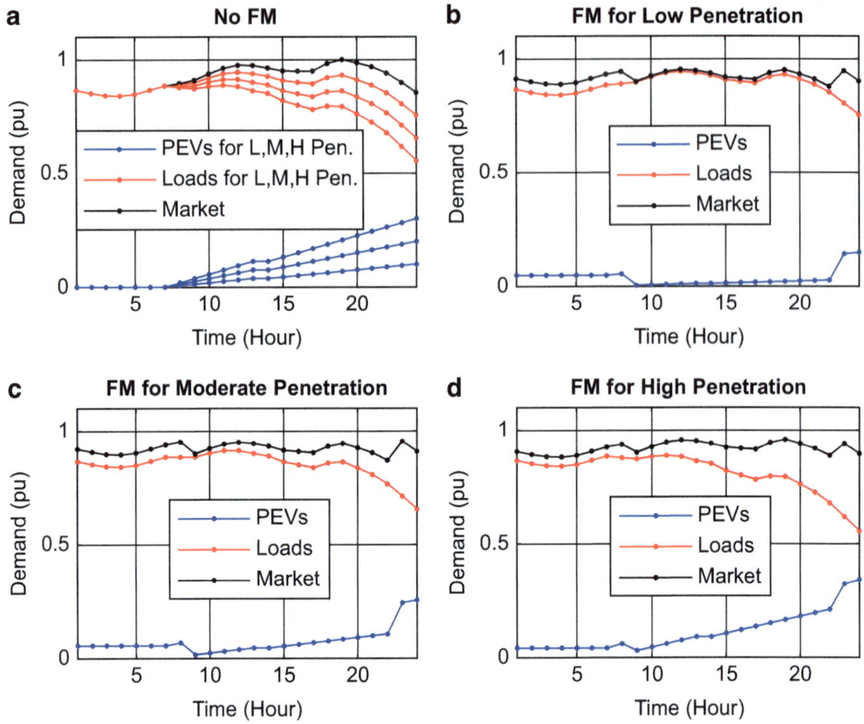

Fig. 1.8 ERCOT power market, moderate-income drivers: (**a**) hourly demand profiles before FM for low (L), moderate (M), and high (H) PEV penetration levels. (**b**) Hourly demand profiles after FM for low PEV penetration (with 70% incentive as the optimal discount). (**c**) Hourly demand profiles after FM for moderate PEV penetration (with 40% incentive as the optimal discount). (**d**) Hourly demand profiles after FM for high PEV penetration (with 20% incentive as the optimal discount)

1.3.3 Sensitivity Analysis with Respect to Social Class of Drivers

In this part, maximum load factor sensitivity analysis is performed with respect to the social class of responsive drivers. In addition, the value of error in the amounts of maximum load factor and optimal incentive, due to the unrealistic modelling of drivers' social class, are determined. As can be seen in Fig. 1.13, changing the social class of drivers alters the maximum achievable load factor of each power market for any PEV penetration level. Moreover, as can be seen in Fig. 1.14, each market needs almost different optimal values of incentive for the social classes of drivers and PEV penetration levels.

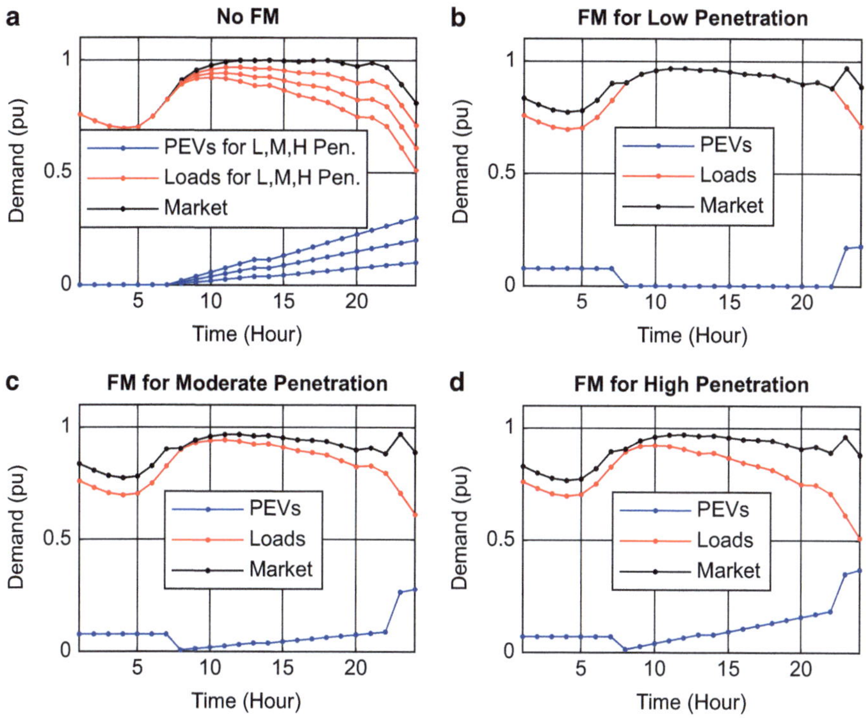

Fig. 1.9 NYISO power market, moderate-income drivers: (**a**) hourly demand profiles before FM for low (L), moderate (M), and high (H) PEV penetration levels. (**b**) Hourly demand profiles after FM for low PEV penetration (with 100% incentive as the optimal discount). (**c**) Hourly demand profiles after FM for moderate PEV penetration (with 50% incentive as the optimal discount). (**d**) Hourly demand profiles after FM for high PEV penetration (with 30% incentive as the optimal discount)

Herein, motivating the high-income drivers to transfer their charging demand from the peak period to the off-peak period needs more incentive compared to the low-income and moderate-income drivers. Furthermore, for the high PEV penetration level, each market operator suggests less incentive to the responsive drivers.

The effects of unrealistic modelling of drivers' social class (considering moderate-income social class instead of realistic social classes) in each power market, considering different PEV penetration levels, are illustrated in Figs. 1.15 and 1.16. Herein, the values of error in the optimal value of discount (%) and in the maximum value of load factor (%) are calculated. As can be seen in Figs. 1.15a and 1.16a, considering the moderate-income social class instead of the realistic low-income and high-income classes misleadingly results in less and more incentives, respectively. Moreover, as can be seen in Figs. 1.15b and 1.16b, the

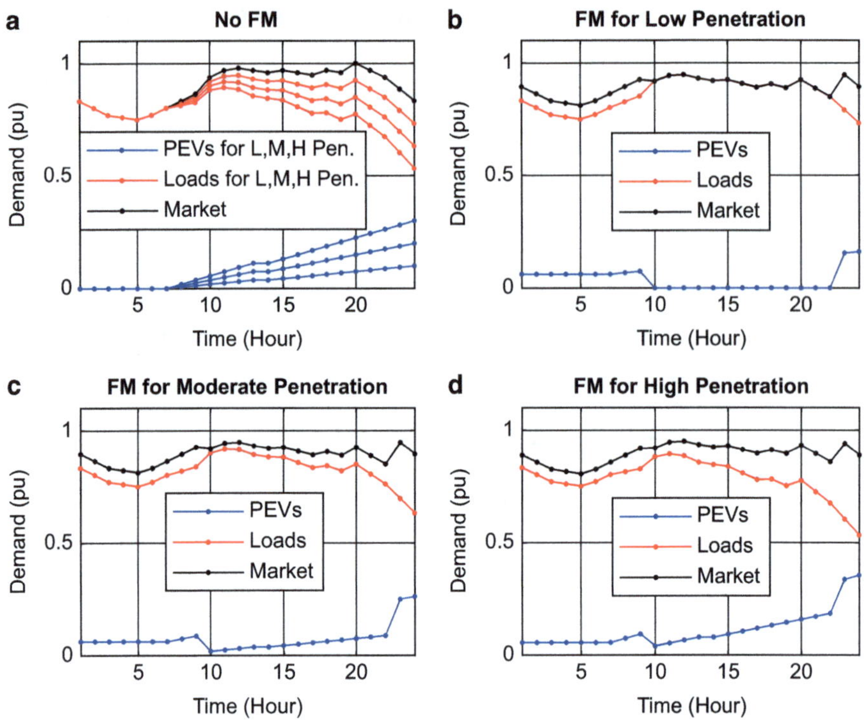

Fig. 1.10 PJM power market, moderate-income drivers: (**a**) hourly demand profiles before FM for low (L), moderate (M), and high (H) PEV penetration levels. (**b**) Hourly demand profiles after FM for low PEV penetration (with 100% incentive as the optimal discount). (**c**) Hourly demand profiles after FM for moderate PEV penetration (with 50% incentive as the optimal discount). (**d**) Hourly demand profiles after FM for high PEV penetration (with 30% incentive as the optimal discount)

non-optimal value of incentive (caused by the unrealistic modelling of drivers' social class) results in an error in the maximum value of load factor of each power market.

1.4 Conclusion

The effects of plug-in electric vehicles' (PEV) charging demand on the demand profile of some of the real power markets were modelled and studied considering the technical and social aspects of the problem. By optimal fleet management (FM) of PEVs, the optimal value of incentive (discount on charging fee) was introduced to the drivers, the responsive drivers transferred part of their charging demand from the peak period to the off-peak one, the demand profile of each market was leveled somewhat, and the related load factor was maximized for any PEV penetration level.

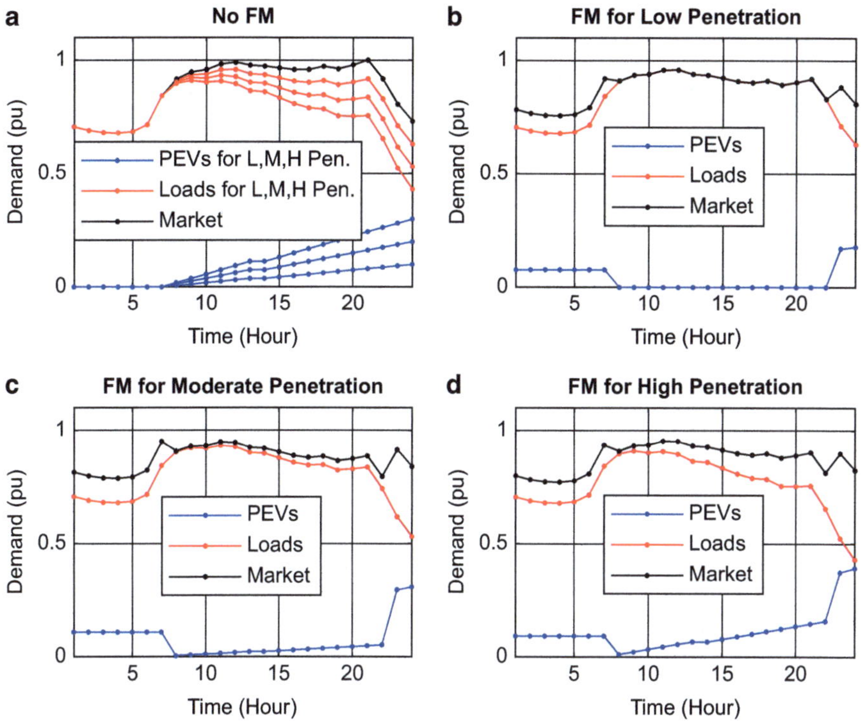

Fig. 1.11 ISO-NE power market, moderate-income drivers: (**a**) hourly demand profiles before FM for low (L), moderate (M), and high (H) PEV penetration levels. (**b**) Hourly demand profiles after FM for low PEV penetration (with 100% incentive as the optimal discount). (**c**) Hourly demand profiles after FM for moderate PEV penetration (with 70% incentive as the optimal discount). (**d**) Hourly demand profiles after FM for high PEV penetration (with 40% incentive as the optimal discount)

It was observed that the optimal value of incentive needs to be investigated in any power market for any PEV penetration level, since each case requires a relatively different optimal value of incentive, and a predetermined incentive is not effective. In addition, it was seen that the optimal value of incentive is almost unique, and in any power market, the case of low PEV penetration is concerned with the highest value of optimal incentive.

It was proven that the social class of drivers changes the optimal value of incentive and the maximum achievable load factor of each power market for any PEV penetration level. In other words, the unrealistic modelling of drivers' social class results in an error in the optimal value of incentive and maximum load factor. Moreover, it was seen that motivating the high-income drivers to transfer their charging demand from the peak period to the off-peak one needs more incentive compared to the other drivers' encouragement.

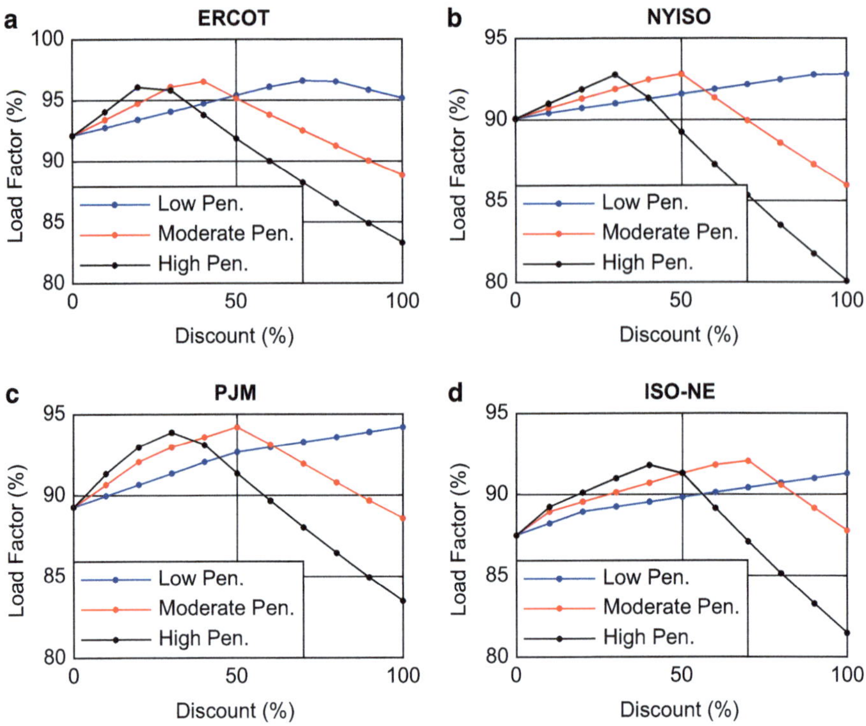

Fig. 1.12 Markets' load factor (with moderate-income drivers) with respect to value of incentive considering different PEV penetration levels in (**a**) ERCOT, (**b**) NYISO, (**c**) PJM, and (**d**) ISO-NE power markets

Therefore, it can be generally concluded that, to have an optimal and cost-effective FM, the social class of drivers must be studied, and an appropriate and optimal incentive scheme needs to be proposed to the responsive drivers.

Nomenclature

Problem parameters	
β	Value of incentive (discount on charging fee of PEVs) in percent
ξ	Responsiveness probability of drivers (%)
μ	Mean of hourly Gaussian distribution function
σ	Standard deviation of hourly Gaussian distribution function
φ^{PEVs}	Penetration percentage of PEVs in the related power market (%)
D	Demand (MW)
$D^{LOADs}(p.u.)$	Loads demand (excluding PEV fleet demand) in per unit
$D^{MARKET}(p.u.)$	Total demand of power market in per unit

(continued)

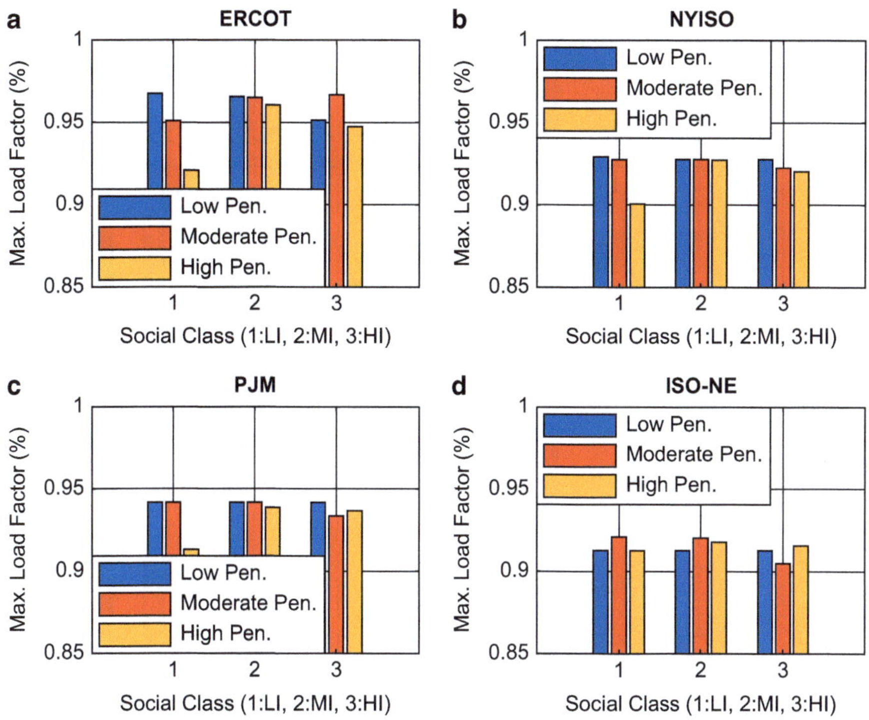

Fig. 1.13 Maximum load factor of market with respect to social classes of drivers (low-income (LI), moderate-income (MI), and high-income (HI)) considering different PEV penetration levels in (**a**) ERCOT, (**b**) NYISO, (**c**) PJM, and (**d**) ISO-NE power markets

$\tilde{D}^{PEVs}$	Charging demand of responsive PEV fleet (MW)
$D^{PEVs,\ TOTAL}$	Total demand of PEVs (including responsive and nonresponsive PEVs) in MW
D^{PEVs}	Charging demand of nonresponsive PEVs (MW)
$\tilde{D}^{PEVs}(p.u.)$	Charging demand of responsive PEV fleet in per unit
$D^{PEVs,\ TOTAL}(p.u.)$	Total demand of PEVs (including responsive and nonresponsive PEVs) in per unit
$D^{PEVs}(p.u.)$	Charging demand of nonresponsive PEVs in per unit
e	Index of PEV
$ERCOT$	Electric Reliability Council of Texas power market
$ISO-NE$	Independent System Operator New England power market
LF	Load factor (%)
LI, MI, HI	Indices of low-income, moderate-income, and high-income drivers
$MARKET$	Index of a power market
$NYISO$	New York Independent System Operator power market

(continued)

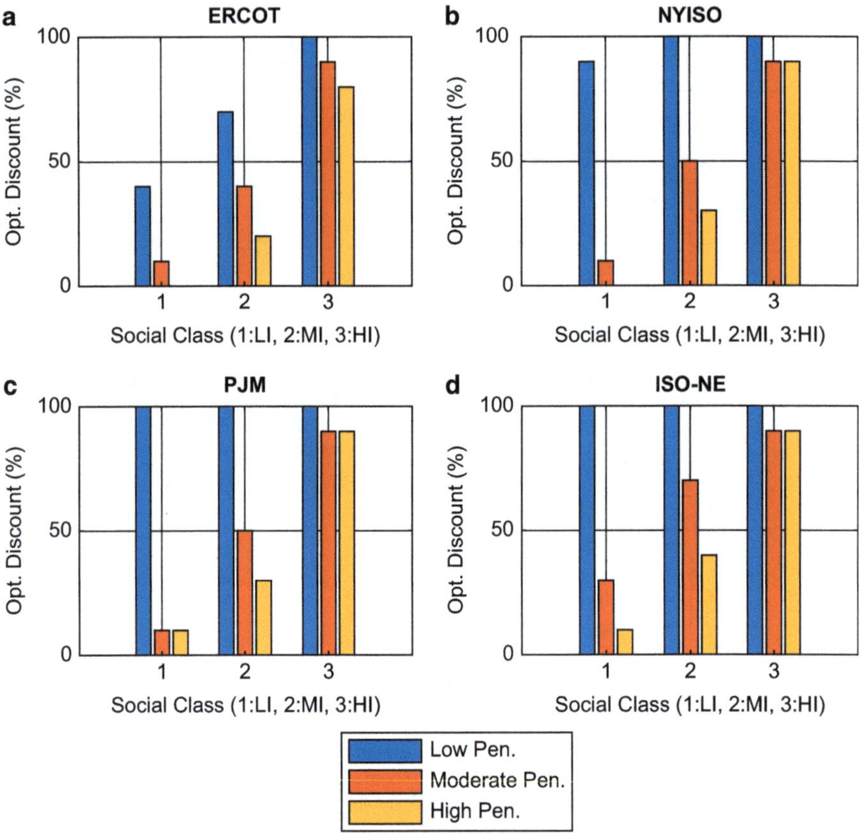

Fig. 1.14 Optimal value of discount for social classes of drivers (low-income (LI), moderate-income (MI), and high-income (HI)) considering different PEV penetration levels in (**a**) ERCOT, (**b**) NYISO, (**c**) PJM, and (**d**) ISO-NE power markets

$\tilde{N}^{PEVs}$	Expected number of responsive drivers
$N^{PEVs,\ TOTAL}$	Total number of PEVs in the area
$Off-peak$	Off-peak period
P^{PEV}	Rated power of a PEV (kW)
Peak	Peak period
PJM	Pennsylvania-Jersey-Maryland power market
SC	Social class of drivers
SOC^{PEVs}	State of charge of PEV fleet (%)
t	Index of time (hour)

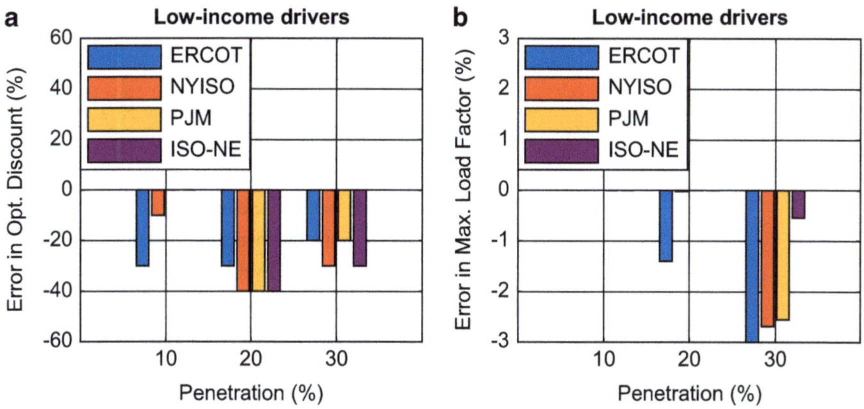

Fig. 1.15 Effects of unrealistic modelling of drivers' social class (moderate-income instead of low-income) in each market considering low (10%), moderate (20%), and high (30%) PEV penetration levels. (**a**) Error in the optimal value of discount (%). (**b**) Error in the maximum value of load factor (%)

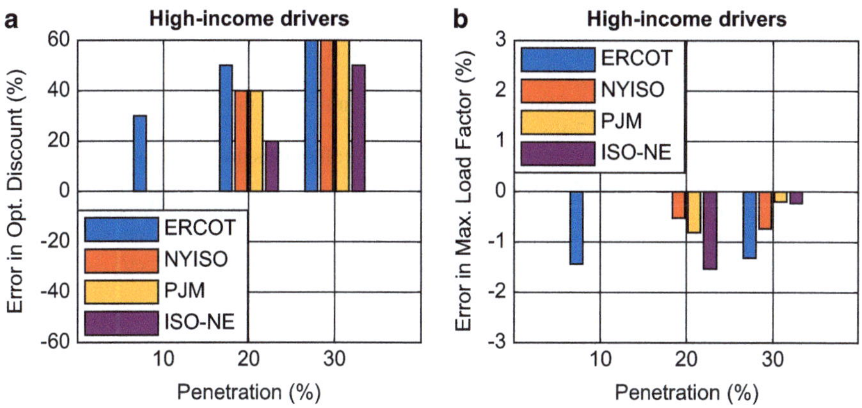

Fig. 1.16 (**a**) Effects of unrealistic modelling of drivers' social class (moderate-income instead of high-income) in each market considering low (10%), moderate (20%), and high (30%) PEV penetration levels. (**a**) Error in the optimal value of discount (%). (**b**) Error in the maximum value of load factor (%)

Abbreviations

DISCO	Distribution company
DR	Demand response
DSM	Demand side management
ERCOT	Electric Reliability Council of Texas
FERC	Federal Energy Regulatory Commission

(continued)

FM	Fleet management
GENCO	Generation company
ISO	Independent system operator
ISO-NE	Independent System Operator New England
LF	Load factor
LI, MI, HI	Low-income, moderate-income, high-income
MPC	Model predictive control
NYISO	New York Independent System Operator
PEV	Plug-in electric vehicle
PJM	Pennsylvania-Jersey-Maryland
PV	Photovoltaic
RES	Renewable energy source
SOC	State of charge

References

1. International Energy Agency (IEA). [Online]. Available: https://iea.org/statistics. Accessed in May 2018
2. [Online]. Available: https://www.sciencenews.org/article/list-diseases-linked-air-pollution-growing. Accessed in May 2018
3. "Monthly plug-in sales scorecard," [Online]. Available: http://insideevs.com/monthly-plug-in-sales-scorecard. Accessed on: Aug 2018
4. Clean Energy Ministerial, "Electric vehicles initiative," [Online]. Available: http://cleanenergyministerial.org/Our-Work/Initiatives/Electric-Vehicles. Accessed on: Aug 2018
5. [Online]. Available: https://www.teslarati.com/map/4362/Atlanta-Supercharger/. Accessed in May 2018
6. [Online]. Available: https://www.pluglesspower.com/learn/driving-electricity-cheaper-gas-50-states/. Accessed in June 2018
7. [Online]. Available: https://www.pluglesspower.com/learn/updated-2017-incentives-electric-vehicles-evse-state-federal-tax-credits-grants-loans-rebates/. Accessed in June 2018
8. Worldwide survey of network-driven demand-side management projects, Task XV, IEA-DSM Res. Rep. No. 1, 2006
9. IEA, "Strategic plan for the IEA demand-side management program 2008–2012" [Online]. Available: http://www.ieadsm.org
10. U.S. Department of Energy, Benefit of demand response in electricity markets and recommendations for achieving them, in: A report to the United States Congress, 2006 [Online]. Available: http://energy.gov/sites/prod/files/oeprod/DocumentsandMedia/DOEBenefitsofDemandResponseinElectricityMarketsandRecommendationsforAchievingThemReporttoCongress.pdf
11. M. Rahmani-Andebili, H. Shen, M. Fotuhi Firuzabad, Planning and operation of parking lots considering system, traffic, and drivers behavioral model. IEEE Trans. Syst. Man Cybern. Syst. (2018). https://doi.org/10.1109/TSMC.2018.2824122
12. M. Rahmani-Andebili, M. Fotuhi Firuzabad, An adaptive approach for PEVs charging management and reconfiguration of electrical distribution system penetrated by renewables. IEEE Trans. Ind. Inf. **14**(5), 2001 (2018)

13. H. Kikusato et al., Electric vehicle charge-discharge management for utilization of photovoltaic by coordination between home and grid energy management systems. IEEE Trans. Smart Grid, 1–12 (2018). https://doi.org/10.1109/TSG.2018.2820026
14. Z. Moghaddam, I. Ahmad, D. Habibi, Q.V. Phung, Smart charging strategy for electric vehicle charging stations. IEEE Trans. Transportation Electrification **4**(1), 76–88 (2018)
15. M. Rahmani-Andebili, M. Fotuhi Firuzabad, M. Moeini-Aghtaie, Optimal incentive plans for plug-in electric vehicles, in *Electric Distribution Network Planning*, (Springer Singapore, 2018), pp. 299–320
16. C. Luo, Y. Huang, V. Gupta, Stochastic dynamic pricing for EV charging stations with renewable integration and energy storage. IEEE Trans. Smart Grid **9**(2), 1494–1505 (2018)
17. A.S. Bin Humayd, K. Bhattacharya, A novel framework for evaluating maximum PEV penetration into distribution systems. IEEE Trans. Smart Grid **9**(4), 2741–2751 (2018)
18. M. Rahmani-Andebili, H. Shen, Traffic and grid-based parking lot allocation for PEVs considering driver behavioral model, in *International Conference on Computing, Networking and Communications (ICNC),* Silicon Valley, USA, 26–29 Jan 2017
19. D.F.R. Melo, A. Trippe, H.B. Gooi, T. Massier, Robust electric vehicle aggregation for ancillary service provision considering battery aging. IEEE Trans. Smart Grid **9**(3), 1728–1738 (2018)
20. Z. Liu, Q. Wu, S.S. Oren, S. Huang, R. Li, L. Cheng, Distribution locational marginal pricing for optimal electric vehicle charging through chance constrained mixed-integer programming. IEEE Trans. Smart Grid **9**(2), 644–654 (2018)
21. M. Rahmani-Andebili, G.K. Venayagamoorthy, Co-operative responsive electric vehicles for social-economic dispatch, in *Cyber-Physical- Social Systems and Constructs in Electric Power Engineering*, (IET, Stevenage, 2016)
22. M. Rahmani-Andebili, G.K. Venayagamoorthy, SmartPark placement and operation for improving system reliability and market participation. Electr. Pow. Syst. Res. **123**(6), 21–30 (2015)
23. M. Rahmani-Andebili, Modelling nonlinear incentive-based and price-based demand response programs and implementing on real power markets. Electr. Pow. Syst. Res. **132**, 115–124 (2016)
24. [Online]. Available: http://www.ferc.gov/market-oversight/mkt-electric/overview.asp. Accessed on Aug 2018
25. D. Newbery, The economics of electric vehicles, in *EPRG and Imperial College London, E&E Seminar*, Cambridge, U.K., Jan 2013. [Online]. Available: http://www.eprg.group.cam.ac.uk/wpcontent/uploads/2013/01/EEJan13_EconomicsEVs.pdf

Chapter 2
Studying the Effects of Optimal Fleet Management of Plug-In Electric Vehicles on the Unit Commitment Problem Considering the Technical and Social Aspects

2.1 Introduction

Plug-in electric vehicles (PEVs) are gaining more popularity in the auto-market [1, 2], since they can be charged by the renewable energy sources (RES) that are the clean and free sources of energy. Figure 2.1 shows the PEV utilization (in million) forecasted by Bloomberg New Energy Finance (BNEF), International Energy Agency (IEA), and some oil and gas industry companies like BP, ExxonMobil, and OPEC, in some different years [3]. As can be noticed, most of the oil producers and research organizations around the world are revising their PEV application forecasts upward due to the decreasing trend of battery cost. In fact, based on the data predicted by the abovementioned organizations and companies, not only the predicted application of PEVs is growing but also the forecasted growth rate is being accelerated. In addition, Fig. 2.2 illustrates the number of PEVs on the road (million) predicted by OPEC in 2015 and 2016 [4]. As can be seen, OPEC has revised its predictions upward in 2016, compared to the ones of 2015. The world PEV and internal combustion engine (ICE) vehicle sales, forecasted by some research organizations, are shown in Fig. 2.3 [5]. As can be seen, it is predicted that the PEV sales will surpass the ICE cars by 2039.

PEVs can act as the energy storage systems and shift part of the system demand from the peak period to the off-peak one to level the demand profile and consequently decrease the commitment of the most expensive and pollutant generation units, resulting in the generation cost reduction.

Fleet management (FM) of PEVs can be considered as one of the demand side management (DSM) measures. DSM is considered as the first precedence in the energy policy decisions due to its benefits from the economic and environmental viewpoints [6]. DSM provides the short-term responses to the electricity market conditions to reduce the overall costs of energy supply, increase reserve margin, mitigate price volatility, and achieve environmental goals by deferring the

M. Rahmani-Andebili, *Planning and Operation of Plug-In Electric Vehicles*,
https://doi.org/10.1007/978-3-030-18022-5_2

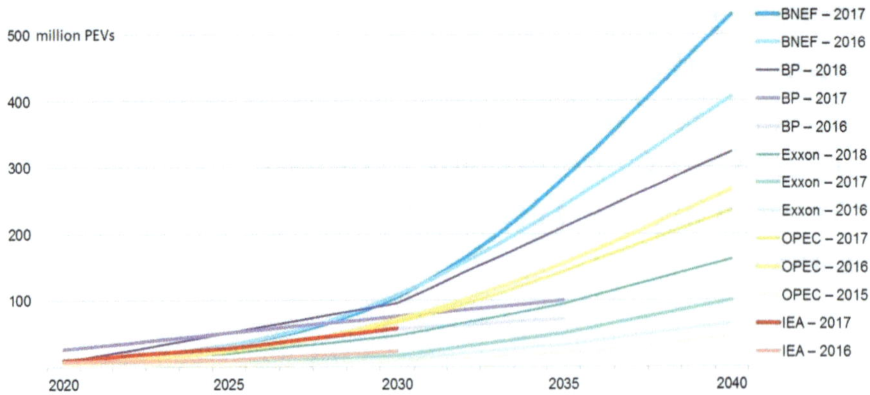

Fig. 2.1 The PEVs utilization (million) forecasted by Bloomberg New Energy Finance (BNEF), International Energy Agency (IEA), and some oil and gas industry companies like BP, ExxonMobil, and OPEC, in some different years [3]

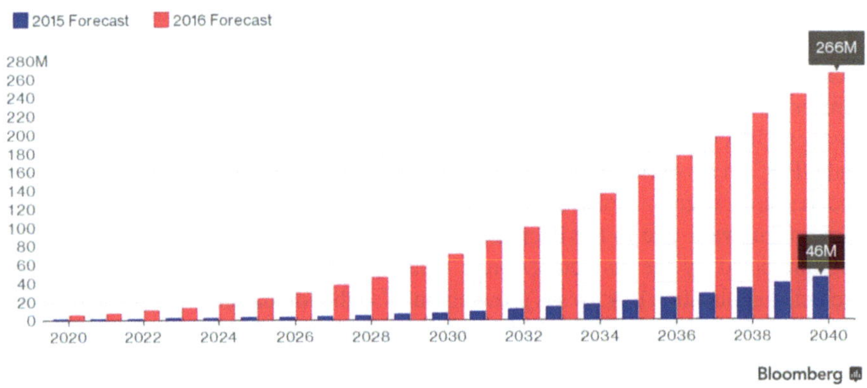

Fig. 2.2 The number of PEVs on the road (million) predicted by OPEC in 2015 and 2016 [4]. (Source: Bloomberg New Energy Finance)

commitment of polluted units leading to the increased energy efficiency and reduced greenhouse gas emissions [7].

According to the Federal Energy Regulatory Commission (FERC), demand response (DR) is defined as the changes in the electric usage by the end-use customers from their normal consumption patterns in response to the changes in the price of electricity over time or to the incentive payments designed to induce lower electricity use at times of high wholesale market prices or when the system reliability is jeopardized [8].

Reference [9] is the first study that has investigated the implementation of DR programs in the unit commitment (UC) problem. In [10, 11], nonlinear DR programs have been studied in the UC problem. In [12–14], carbon emission of generation units has been modelled in the UC problem. In [15], a non-conservative security-

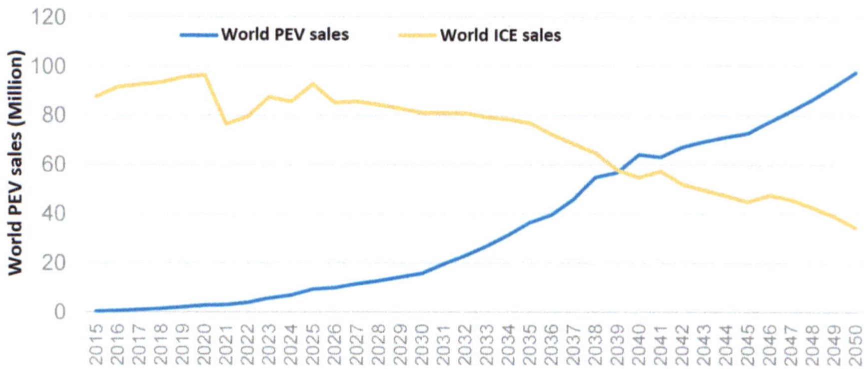

Fig. 2.3 The world PEV and internal combustion engine (ICE) vehicles sales predicted by some research organizations [5]. (Source: Ward's, ACEA, CAAM, Morgan Stanley Research estimates (from 2017 onward))

constraint UC model approach has been proposed. Reference [16] has studied a Security-constrained unit commitment (SCUC) problem (N-1 criterion) based on the line outage distribution factors. In [17], an optimization approach for the contingency-constrained UC problem, based on the N-K criterion, has been presented. In [18], DR programs have been studied in a risk-cost-based UC problem. In [19], a clustered UC formulation has been proposed to provide the accurate and fast approximations to the UC problem.

In [20], the impact of PEVs integration on the power system has been studied. Reference [21, 22] has performed a study on the integration of generation system and responsive PEVs. The main shortcoming of the abovementioned studies is disregarding the social aspect of the problem and the drivers' behavioral model in the generation scheduling and UC problem. The contributions of chapter include:

- Modelling and studying the effects of PEVs on the generation scheduling and UC problem
- Implementing the optimal FM of PEVs considering different PEV penetration levels
- Categorizing the drivers in three different social classes based on their income level
- Modelling the behavior of each social class of drivers based on their reaction with respect to the value of incentive, suggested by the generation company (GENCO) to the drivers, to transfer their charging demand from the peak period to the off-peak period
- Performing the sensitivity analysis for the total cost of problem with respect to value of incentive considering different PEV penetration levels and various social classes of drivers
- Studying the effects of unrealistic modelling of drivers' social class on the optimal value of incentive, minimum total cost of problem, and generation scheduling and commitment of generation units

In the following, in Sect. 2.2, the proposed approach that includes the social and technical aspects of problem is presented and described. In Sect. 2.3, the optimization technique is presented. Section 2.4 deals with the problem formulation. In Sect. 2.5, the problem is simulated, and the results are analyzed. Finally, Sect. 2.6 concludes the chapter.

2.2 Modelling the Social and Technical Aspects of Problem

2.2.1 Social Classification of Drivers

In this study, the drivers are stratified in three different social classes based on their income level, namely, low-income, moderate-income, and high-income. From the sociology and economic points of view, this type of classification is necessary, since the drivers, with different social classes, will differently react with respect to the value of incentive suggested by a GENCO. In other words, it is necessary to model the willingness or reluctance level of different social classes of drivers, since it is obvious that the low-income drivers are more likely to be interested in the incentive compared to the high-income drivers. Herein, the responsiveness probability of each social class of drivers is modeled with respect to the value of incentive (β) based on the following logics:

- The probability that a driver (with any social class) will transfer his/her charging demand from the peak period to the off-peak one is 0% if there is no incentive ($\beta = 0\%$).
- The probability that a driver (with any social class) will transfer his/her charging demand from the peak period to the off-peak period is 100% if the PEV is charged in free ($\beta = 100\%$).

Since the low-income and high-income drivers are more and less likely to be interested in the proposed incentive, a power function and an exponential function are applied to model their behaviors, respectively. A power function and an exponential function return large and small values for the given small value of variable, respectively. Therefore, by considering the abovementioned criteria and the proposed power and exponential functions, the responsiveness probability functions of low-income (LI) and high-income (HI) drivers can be formulated as Eqs. (2.1) and (2.2). In addition, a linear function is used to model the behavior of moderate-income (MI) drivers, as can be seen in Eq. (2.3). Herein, t and e are the indices of time and PEV, respectively. The functions determine the responsiveness probability of a driver to transfer his/her charging demand from the peak period to the off-peak one while being called by the GENCO.

Figure 2.4 illustrates the bar chart of responsiveness probability of low-income, moderate-income, and high-income drivers with respect to the value of discount on charging fee (β, as the incentive) in percent. The expected number of responsive drivers ($\tilde{N}^{PEVs}$) that will transfer their charging demand from the peak period to the

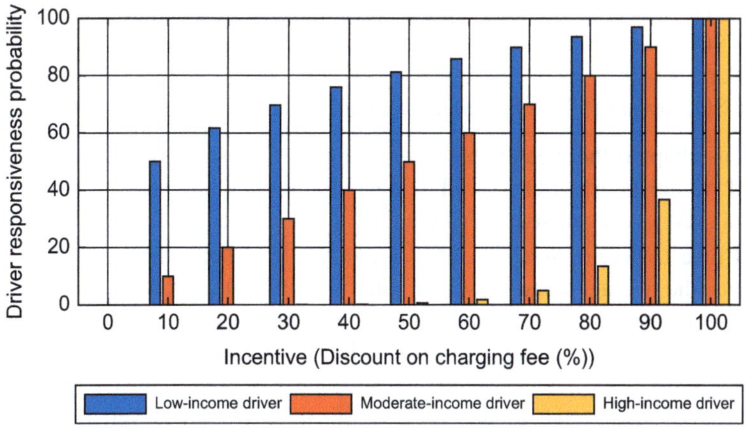

Fig. 2.4 Responsiveness probability of a low-income, moderate-income, and high-income driver with respect to the value of discount on charging fee (β) for $\beta \in \{0\%, 10\%, \ldots, 100\%\}$

off-peak period is determined using Eq. (2.4). Herein, SC and $N^{PEVs,\ TOTAL}$ are the index of social class and total number of PEVs in the area, respectively.

$$\xi_{e,LI,\beta,t} = 100 \times \left(\frac{\beta}{100}\right)^{0.3} \tag{2.1}$$

$$\xi_{e,HI,\beta,t} = 100 \times exp\left(10 \times \left(\frac{\beta}{100} - 1\right)\right) \tag{2.2}$$

$$\xi_{e,MI,\beta,t} = 100 \times \left(\frac{\beta}{100}\right) \tag{2.3}$$

$$\tilde{N}^{PEVs}_{SC,\beta,t} = \sum_{e=1}^{N^{PEVs,TOTAL}} \frac{\xi_{e,SC,\beta,t}}{100}, \ \forall SC \in \{LI, MI, HI\},\ \forall \beta \in \{0, 10, \ldots, 100\},\ \forall t \in \{1, \ldots, 24\} \tag{2.4}$$

2.2.2 Technical Aspect

In [23], a stochastic model predictive control (MPC) approach has been applied to model the variability and uncertainty of state of charge (SOC) of PEV fleet. In this reference, the hourly SOC of fleet has been modelled by the mean (μ) and standard deviation (σ) of hourly Gaussian distribution function. In the modelling, some facts can be identified as follow:

- The SOC level of PEV fleet has a descending trend throughout the day.
- The SOC level of PEV fleet has smaller standard deviations during the periods of 1–7 and 22–24.
- The SOC level of PEVs has larger standard deviations during the period of 8–21.

In this chapter, the mean value of each hourly Gaussian distribution function is applied as the hourly SOC of PEV fleet. Equation (2.5) presents the charging demand (MW) of responsive PEV fleet ($\tilde{D}^{PEVs}_{SC,\beta,t}$) with social class of SC and discount of β at time t. Herein, SOC^{PEVs}, P^{PEVs}, and $\tilde{N}^{PEVs}$ indicate the SOC of PEV fleet (percent), the average rated power of battery of PEV fleet (kW), and the expected number of responsive drivers that will transfer their charging demand from the peak period to the off-peak one, respectively. In Eq. (2.6), $D^{PEVs,\ TOTAL}$ presents the total demand (MW) of PEV fleet (including responsive and nonresponsive PEVs), where $N^{PEVs,\ TOTAL}$ is the total number of PEVs in the area. Moreover, D^{PEVs} in (2.7) represents the hourly charging demand of nonresponsive PEV fleet (MW). Demand of other loads (D^{LOADs}) is determined using Eq. (2.8), where D^{TOTAL} (MW) is the total demand of system. The hourly demands of PEV fleet are calculated using their hourly SOC level (presented in [23]) and assuming Nissan Leaf as the type of PEVs with the rated power and capacity of 24 kW and 24 kWh, respectively [24].

$$\tilde{D}^{PEVs}_{SC,\beta,t} = \left(1 - \frac{SOC_t^{PEVs}}{100}\right) \times \frac{P^{PEVs}}{1000} \times \tilde{N}^{PEVs}_{SC,\beta,t} \tag{2.5}$$

$$D_t^{PEVs,TOTAL} = \left(1 - \frac{SOC_t^{PEVs}}{100}\right) \times \frac{P^{PEVs}}{1000} \times N^{PEVs,TOTAL} \tag{2.6}$$

$$D_t^{PEVs} = D_t^{PEVs,TOTAL} - \tilde{D}^{PEVs}_{SC,\beta,t} \tag{2.7}$$

$$D_t^{LOADs} = D_t^{TOTAL} - D_t^{PEVs,TOTAL} \tag{2.8}$$

2.3 Optimization Technique

In this section, genetic algorithm (GA) is applied as the optimization technique to solve the UC problem. The inverse value of total cost of problem is defined as the fitness of a chromosome, and then the GA tries to minimize the fitness of chromosomes. A chromosome, illustrated in Fig. 2.5, represents the hourly status of generation units in a day.

The problem is solved for each possible value of incentive (discount on charging fee of PEVs) with a 10% step, that is, 0%, 10%, ..., 100%. Then, the optimal value of incentive is determined based on the minimum total cost of problem. In the following, the steps for applying GA in the optimization problem are presented and described.

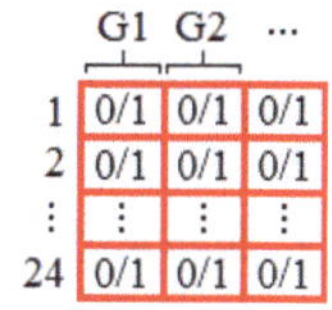

Fig. 2.5 The structure of a chromosome

- *Step 1: Obtaining the primary data*

Parameters for applying GA: These parameters include the mutation probability of genes (θ^M) and the number of the chromosomes (N^C) as the size of population.

Parameters of system: The value of all the parameters of system and problem are obtained. Also, the value of incentive (β as value of discount on charging fee) is chosen.

Updating the PEVs charging demand and the system demand: The charging demand of responsive drivers is updated based on the value of incentive, and then the system demand is modified.

Initial population: The chromosomes of population are initialized with random binary values ("0" or "1").

- *Step 2: Updating the population*

Applying crossover operator: The crossover operator is applied on every two chromosomes to reproduce two new chromosomes as the offspring.

Applying mutation operator: The mutation is applied on every gene of every chromosome of population with the definite probability of θ^M.

- *Step 3: Selecting the new population*

Evaluating the fitness of every chromosome: For every chromosome, the optimal generation scheduling of units is determined using Lambda-Iteration Economic Dispatch method [25], and if all the constraints are satisfied, the fitness (f_c) of chromosome is calculated.

Applying selection process: The new chromosomes are selected using the probabilistic fitness-based selection (PFBS) technique, where the fitter chromosomes are more likely to be chosen. Herein, r_c is a random number between [0,100] generated for the chromosome (c), and a_c is the acceptance indicator of a chromosome for the new population.

$$a_c = \begin{cases} 1 & \theta_c^{PFBS} > r_c \\ 0 & \theta_c^{PFBS} < r_c \end{cases} \tag{2.9}$$

The value of selection probability of a chromosome (θ_c^{PFBS}) is determined using Eq. (2.10), which is proportional to the fitness of chromosome.

$$\theta_c^{PFBS} = \frac{f_c}{Max\{f_i | i \in \{1, \ldots, N^C\}\}} \times 100 \tag{2.10}$$

- *Step 4: Checking the termination criterion*
  In this step, the convergence status of optimization procedure is checked. Based on this, the value of improvement in the fitness of chromosomes of old and new populations is computed, and if there is no significant improvement in them, the optimization process is terminated; otherwise, the algorithm is continued from Step 2.
- *Step 5: Introducing the outcomes*
  The consequences include the minimum total cost of problem, the optimal generation scheduling and commitment of generation units, and the updated demands of PEVs and system.

2.4 Problem Formulation

In this section, the mathematical formulation for the operation problem of GENCO is presented. The goal of GENCO is minimizing total daily operation cost of generation system. In addition, the problem outputs include the optimal value of incentive, the optimal generation scheduling and commitment of generation units, and the modified demands of PEVs and system.

2.4.1 *Objective Function of Problem*

The objective function of problem is presented in Eq. (2.11) that includes some cost terms such as cost of discount on charging fee of PEVs ($Cost^{INC}$), fuel cost of generation units ($Cost^{F}$), greenhouse gas emissions cost of generation units ($Cost^{E}$), start-up cost of decommitted units ($Cost^{STU}$), and shut-down cost of committed units ($Cost^{SHD}$). Herein, g and N^G are the index of generation unit and total number of generation units, respectively.

$$OF = min\left\{ \sum_{t=1}^{24} \left[Cost_{SC,\beta,t}^{INC} + \sum_{g=1}^{N^G} \left[Cost_{g,t}^{F} + Cost_{g,t}^{E} + Cost_{g,t}^{STU} + Cost_{g,t}^{SHD} \right] \right] \right\} \tag{2.11}$$

2.4.2 Cost Terms of Problem

In the following, the cost terms of objective function are described. All the cost terms are in US dollar.

Incentive cost The incentive cost ($Cost^{INC}$) is the function of demand of responsive PEVs (MW), the value of discount on the charging fee (β) in percent, and the electricity price (π^E) in \$/MWh.

$$Cost_{SC,\beta,t}^{INC} = \tilde{D}_{SC,\beta,t}^{PEVs} \times \frac{\beta}{100} \times \pi^E \tag{2.12}$$

Fuel cost of generation units The fuel cost of a generation unit ($Cost^F$) is a quadratic polynomial of its power (P^G) in MW [25]. The α_1^F, α_2^F, and α_3^F are the fuel cost coefficients of generation unit in \$/MWh^2, \$/MWh, and \$, respectively.

$$Cost_{g,t}^F = \alpha_{1,g}^F \times \left(P_{g,t}^G\right)^2 + \alpha_{2,g}^F \times \left(P_{g,t}^G\right) + \alpha_{3,g}^F \tag{2.13}$$

Greenhouse gas emissions cost of generation units The greenhouse gas emissions cost of a generation unit is a quadratic polynomial of its power (P^G) [25]. The α_1^E, α_2^E, and α_3^E are the emission coefficients of generation unit in Ton/MWh^2, Ton/MW, and Ton, respectively. Also, γ^E is the emission penalty factor in \$/Ton.

$$Cost_{g,t}^E = \gamma^E \times \left(\alpha_{1,g}^E \times \left(P_{g,t}^G\right)^2 + \alpha_{2,g}^E \times \left(P_{g,t}^G\right) + \alpha_{3,g}^E\right) \tag{2.14}$$

Start-up cost and shut-down cost of generation units The start-up cost of a decommitted unit ($Cost^{STU}$) and shut-down cost of a committed unit ($Cost^{SHD}$) are presented in Eqs. (2.15) and (2.16), respectively. Herein, s^G indicates the status of a generation unit, where "1" and "0" mean "on" and "off," respectively. Moreover, C^{STU} and C^{SHD} are the start-up and shut-down costs, respectively.

$$Cost_{g,t}^{STU} = C_g^{STU} \times \left(1 - s_{g,t-1}^G\right) \times s_{g,t}^G \tag{2.15}$$

$$Cost_{g,t}^{SHD} = C_g^{SHD} \times s_{g,t-1}^G \times \left(1 - s_{g,t}^G\right) \tag{2.16}$$

2.4.3 Constraints of Problem

All the constraints must be held at each time step (hour) of operation period (1 day).

System power balance constraint This constraint is presented in Eq. (2.17).

$$\sum_{g=1}^{N^G} P_{g,t}^{G} \times s_{g,t}^{G} = D_t^{LOADs} + D_t^{PEVs} + \tilde{D}_{SC,\beta,t}^{PEVs} \tag{2.17}$$

Minimum generation constraint The generation units, which are "on," must be able to supply the minimum demand of system, as can be seen in Eq. (2.18).

$$\sum_{g=1}^{N^G} P_{g}^{G,MIN} \times s_{g,t}^{G} \leq D_t^{LOADs} + D_t^{PEVs} + \tilde{D}_{SC,\beta,t}^{PEVs} \tag{2.18}$$

Maximum generation constraint considering spinning reserve The generation units, which are "on," must be able to supply the maximum demand of system considering the minimum required spinning reserve (*SR*), as can be seen in Eq. (2.19).

$$\sum_{g=1}^{N^G} P_{g}^{G,MAX} \times s_{g,t}^{G} \geq D_t^{LOADs} + D_t^{PEVs} + \tilde{D}_{SC,\beta,t}^{PEVs} + SR_t^{MIN} \tag{2.19}$$

Power constraints of generation units The maximum and minimum power constraints of a generation unit at every hour of operation period are presented in Eq. (2.20).

$$P_{g}^{G,MIN} \leq P_{g,t}^{G} \leq P_{g}^{G,MAX} \tag{2.20}$$

Ramp-up rate and ramp-down rate constraints of generation units The ramp-up rate (RUR^G) and ramp-down rate (RDR^G) constraints of a generation unit at every hour of operation period are presented in Eqs. (2.21) and (2.22), respectively.

$$\left(P_{g,t+1}^{G} - P_{g,t}^{G}\right) \leq RUR_g^G \tag{2.21}$$

$$\left(P_{g,t}^{G}-P_{g,t+1}^{G}\right)\leq RDR_{g}^{G} \tag{2.22}$$

Minimum "down" time and minimum "up" time constraints of generation units The minimum "down" time (MDT^G) and minimum "up" time (MUT^G) constraints of a generation unit at every hour of operation period are presented in Eqs. (2.23) and (2.24), respectively. Herein, $OFFT^G$ and ONT^G indicate the intervals that the generation unit have been "off" and "on," respectively.

$$OFFT_{g,t}^{G}\geq MDT_{g}^{G} \tag{2.23}$$

$$ONT_{g,t}^{G}\geq MUT_{g}^{G} \tag{2.24}$$

2.5 Simulation Results

2.5.1 *Base Case*

The technical data of generation system that includes ten generation units is presented in Table 2.1. The minimum required spinning reserve of system is assumed about 10% of the total demand at each hour. In addition, the value of penalty for the greenhouse gas emissions is considered about $10/Ton based on the California Air Resources Board auction of greenhouse gas emissions [26]. Figure 2.6a illustrates the hourly demand profile of system (MW) that its per unit value is the same as the demand profile of Electric Reliability Council of Texas (ERCOT) power market [27]. Herein, the demand profile above 0.9 p.u. is considered as the peak period (hours: 9–22). Moreover, Fig. 2.6b shows the hourly marginal cost price ($/MWh) calculated based on the results of generation scheduling and UC problem of system before FM. The hourly marginal cost price is considered as the hourly electricity price and charging fee of PEVs.

In this study, the problem is solved for different social classes of drivers, namely, low-income, moderate-income, and high-income, as well as for different PEV penetration levels, that is, low, moderate, and high penetrations. Table 2.2 presents the number of PEVs in each penetration level.

2.5.1.1 Low-Income Drivers

Table 2.3 presents the optimal value of incentive and minimum total cost of problem before and after optimal FM of low-income drivers for the low, moderate, and high PEV penetration levels. As can be seen, the minimum total cost of problem is about 0.55050 million $/day before FM of PEV fleet. After optimal FM of PEV fleet, the

Table 2.1 The technical characteristics of generation units

Generation unit	G1	G2	G3	G4	G5
α_1^F (/MWh^2)	0.00048	0.00031	0.00200	0.00211	0.00398
α_2^F (/MWh)	16.19	17.26	16.60	16.50	19.70
α_3^F ($)	1000	970	700	680	450
α_1^E (Ton/MWh^2)	0.0005	0.0005	0.0005	0.0005	0.0010
α_2^E (Ton/MWh)	0.4050	0.4320	0.4150	0.4120	0.4930
α_3^E (Ton)	0.3000	0.4250	0.4500	0.7000	0.7250
P^{MIN} (MW)	75	75	15	15	15
P^{MAX} (MW)	200	200	120	100	100
MUT (h)	5	5	5	5	5
MDT (h)	−5	−5	−5	−5	−5
RUR (MW/h)	110	110	80	80	80
RDR (MW/h)	110	110	80	80	80
C^{STU} ($)	4500	5000	550	560	900
C^{SHD} ($)	4500	5000	550	560	900
Initial status	+24	+24	+24	+24	+24
Generation unit	**G6**	**G7**	**G8**	**G9**	**G10**
α_1^F (/MWh^2)	0.00712	0.00790	0.00813	0.00822	0.00873
α_2^F (/MWh)	22.26	27.74	25.92	27.27	27.79
α_3^F ($)	370	480	660	665	670
α_1^E (Ton/MWh^2)	0.0020	0.0020	0.0024	0.0025	0.0025
α_2^E (Ton/MWh)	0.5560	1.0940	1.6480	1.6820	1.6950
α_3^E (Ton)	0.9250	1.2000	1.6500	1.6625	1.7750
P^{MIN} (MW)	10	10	10	10	10
P^{MAX} (MW)	80	50	25	20	20
MUT (h)	3	1	1	1	1
MDT (h)	−3	−1	−1	−1	−1
RUR (MW/h)	60	20	10	10	10
RDR (MW/h)	60	20	10	10	10
C^{STU} ($)	170	30	30	30	30
C^{SHD} ($)	170	30	30	30	30
Initial status	−3	−1	−1	−1	−1

minimum total cost of problem decreases for each PEV penetration level; however, high penetration level results in the most cost saving ($7089/day). As can be seen, a special optimal incentive is needed for each PEV penetration level that leads to a different discount cost.

Figure 2.7 shows the PEVs' charging demand, loads demand, and total demand of system, before and after optimal FM of low-income drivers (20% discount), considering the low penetration level for the PEVs. As can be seen, the responsive drivers have shifted part of their charging demand from the peak period to the off-peak period that causes the same pattern for the total demand. Figures 2.8 and 2.9 illustrate

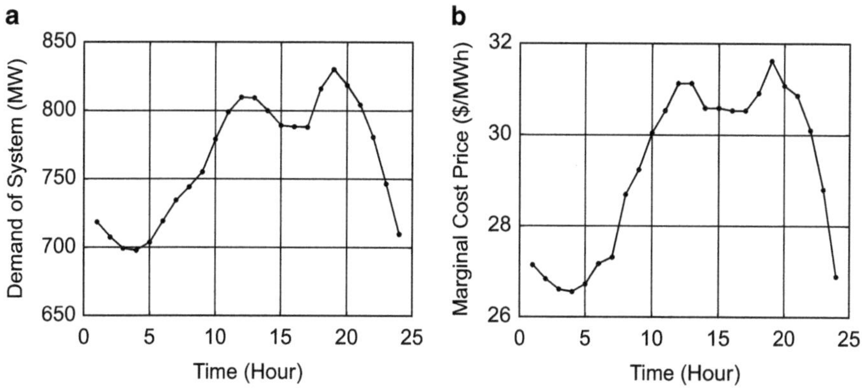

Fig. 2.6 (**a**) Demand profile of system (MW) [27]. (**b**) Marginal cost price ($/MWh) as the electricity price and charging fee of PEVs

Table 2.2 Number of PEVs in each penetration level

Penetration level	Low	Moderate	High
Number of PEVs	1000	3000	5000

Table 2.3 Optimal value of incentive and minimum total cost of problem before and after optimal FM of low-income drivers considering different PEV penetrations

	Before FM	After FM		
Outputs	Any penetration	Low penetration	Moderate penetration	High penetration
Optimal incentive (%)	0	20	30	10
Discount cost ($)	0	444	2258	902
UC cost (million $/day)	0.55050	0.54699	0.54147	0.54250
Minimum total cost (million $/day)	0.55050	0.54743	0.54373	0.54341
Cost saving ($/day)	–	3062	6766	7089

the demands profile, before and after optimal FM of PEVs, considering the moderate penetration level (30% discount) and high penetration level (10% discount). As can be seen, more demand is transferred from the peak period to the off-peak period, and the profile of total demand becomes flatter.

Table 2.4 presents the optimal generation scheduling and commitment of generation units after optimal FM of low-income drivers considering a high PEV penetration level. The values in red color and the highlighted boxes show the differences in the generation scheduling and commitment of units compared to the problem results before FM, respectively. As can be seen, the generation scheduling of units G5–G10 and the commitment of units G7–G10 have changed after optimal FM of PEVs. In fact, the most expensive and pollutant generation units (G8–G10) have been shut down, or their generation level have reduced during the peak period, while

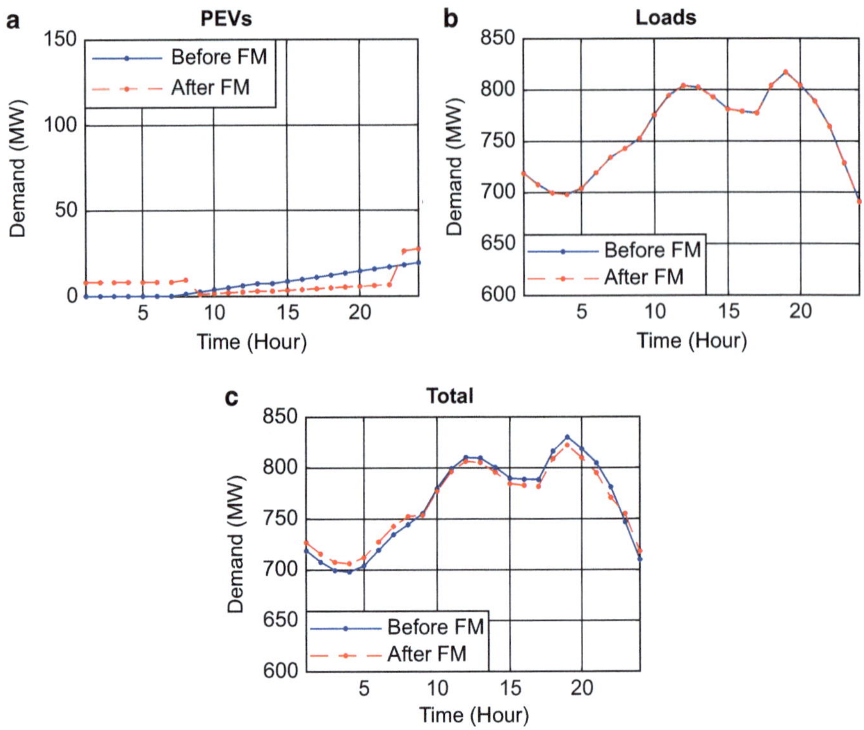

Fig. 2.7 Demands profile before and after optimal FM of low-income drivers (20% discount) considering the low penetration level for the PEVs. (**a**) PEVs' charging demand. (**b**) Loads demand. (**c**) Total demand

the commitment and generation level of the least expensive and the least pollutant units (G5–G7) have increased during the off-peak period.

2.5.1.2 Moderate-Income Drivers

Table 2.5 presents the optimal results of problem before and after optimal FM of moderate-income drivers for different PEV penetration levels. By comparing Tables 2.5 and 2.3, it is noted that the value of cost saving is lower for the FM of moderate-income drivers, for any PEV penetration level. However, like the FM of low-income drivers, the FM of moderate-income drivers results in the most cost saving for the high PEV penetration level ($5320/day). Additionally, the optimal values of incentive for different PEV penetration levels are different from the ones of low-income drivers.

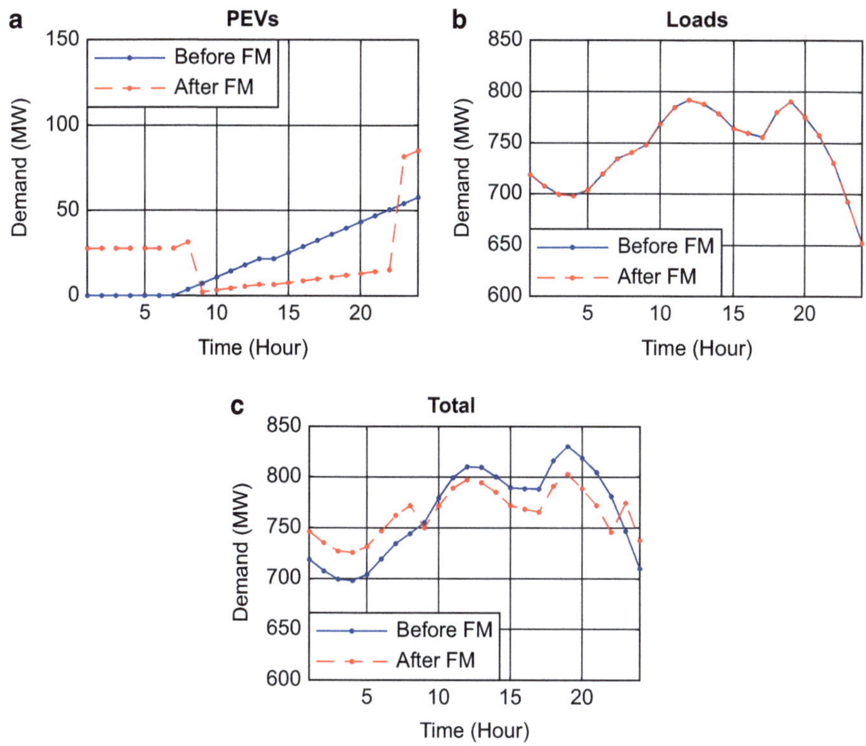

Fig. 2.8 Demands profile before and after optimal FM of low-income drivers (30% discount) considering a moderate penetration level for the PEVs. (**a**) PEVs' charging demand. (**b**) Loads demand. (**c**) Total demand

2.5.1.3 High-Income Drivers

By comparing Tables 2.6, 2.5, and 2.3, it is realized that the FM of high-income drivers needs the largest incentive but results in the least cost saving for the GENCO. As can be seen in Table 2.6, the FM of high-income drivers leads to a relatively slight cost saving for the GENCO, even with high PEV penetration level ($1444/ day).

Table 2.7 gives the optimal generation scheduling and commitment of generation units after optimal FM of high-income drivers considering the high PEV penetration level. The values in red color and the highlighted boxes indicate the differences in the generation scheduling and commitment of units compared to the problem outcomes before FM, respectively. By comparing Tables 2.7 and 2.4, it is noticed that the FM of high-income drivers results in dissimilar generation scheduling and UC.

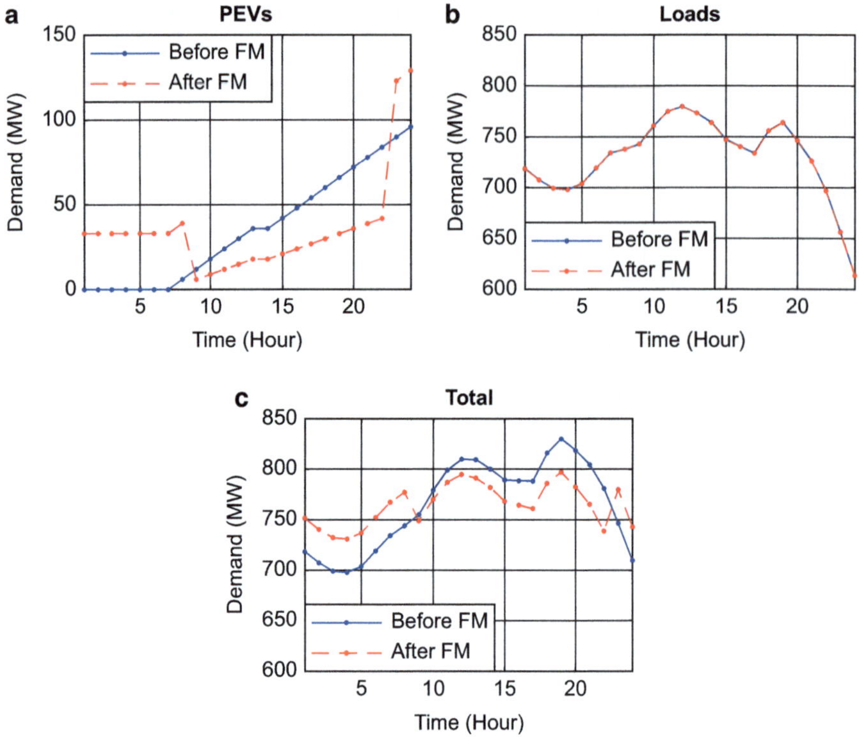

Fig. 2.9 Demands profile before and after optimal FM of low-income drivers (10% discount) considering a high penetration level for the PEVs. (**a**) PEVs' charging demand. (**b**) Loads demand. (**c**) Total demand

2.5.2 Sensitivity Analysis for Operation Cost

Figure 2.10 illustrates the sensitivity analysis for the total cost of problem with respect to the value of incentive considering different PEV penetration levels and social classes of drivers. As can be seen, the curves are intensively nonlinear; however, there is just one unique optimal point for each curve. Moreover, by comparing Fig. 2.10a and c, it is noted that the FM of low-income drivers and high-income drivers have the highest and lowest potentials to minimize the total cost of problem, respectively.

Figure 2.11 shows the sensitivity analysis for the value of discount cost, UC cost, and total cost with respect to the value of incentive (discount on charging fee) concerned with the FM of low-income drivers considering the high PEV penetration level. As can be seen, although the discount cost and UC cost have ascending and uncertain trends with respect to the value of incentive, respectively, the total cost has a specific pattern with one minimum point.

Table 2.4 Optimal generation scheduling and commitment of generation units after optimal FM of low-income drivers considering a high penetration level for the PEVs

Hour	G1	G2	G3	G4	G5	G6	G7	G8	G9	G10
1	200	200	120	100	100	23	10	0	0	0
2	200	200	120	100	100	12	10	0	0	0
3	200	200	120	100	94	10	10	0	0	0
4	200	200	120	100	92	10	10	0	0	0
5	200	200	120	100	98	10	10	0	0	0
6	200	200	120	100	100	24	10	0	0	0
7	200	200	120	100	100	39	10	0	0	0
8	200	200	120	100	100	39	10	10	0	0
9	200	200	120	100	100	20	10	0	0	0
10	200	200	120	100	100	42	10	0	0	0
11	200	200	120	100	100	48	10	10	0	0
12	200	200	120	100	100	56	10	10	0	0
13	200	200	120	100	100	53	10	10	0	0
14	200	200	120	100	100	43	10	10	0	0
15	200	200	120	100	100	40	10	0	0	0
16	200	200	120	100	100	36	10	0	0	0
17	200	200	120	100	100	32	10	0	0	0
18	200	200	120	100	100	47	10	10	0	0
19	200	200	120	100	100	48	10	10	10	0
20	200	200	120	100	100	44	10	10	0	0
21	200	200	120	100	100	37	10	0	0	0
22	200	200	120	100	100	10	10	0	0	0
23	200	200	120	100	100	41	10	10	0	0
24	200	200	120	100	100	14	10	0	0	0

Table 2.5 Optimal value of incentive and minimum total cost of problem before and after optimal FM of moderate-income drivers considering different PEV penetration levels

	Before FM	After FM		
Outputs	Any penetration	Low penetration	Moderate penetration	High penetration
Optimal incentive (%)	0	60	70	40
Discount cost ($)	0	1296	5292	2880
UC cost (million $/day)	0.55050	0.54703	0.54150	0.54229
Minimum total cost (million $/day)	0.55050	0.54832	0.54679	0.54517
Cost saving ($/day)	–	2173	3703	5320

Table 2.6 Optimal value of incentive and minimum total cost of problem before and after optimal FM of high-income drivers considering different penetrations of PEVs

Outputs	Before FM	After FM		
	Any penetration	Low penetration	Moderate penetration	High penetration
Optimal incentive (%)	0	90	80	90
Discount cost ($)	0	1192	1669	5958
UC cost (million $/day)	0.55050	0.54877	0.54870	0.54309
Minimum total cost (million $/day)	0.55050	0.54997	0.54987	0.54905
Cost saving ($/day)	–	529	628	1444

Table 2.7 Optimal generation scheduling and commitment of generation units after optimal FM of high-income drivers considering the high PEV penetration level

Hour	G1	G2	G3	G4	G5	G6	G7	G8	G9	G10
1	200	200	120	100	100	14	10	0	0	0
2	200	200	120	100	93	10	10	0	0	0
3	200	200	120	100	95	10	0	0	0	0
4	200	200	120	100	94	10	0	0	0	0
5	200	200	120	100	90	10	10	0	0	0
6	200	200	120	100	100	15	10	0	0	0
7	200	200	120	100	100	30	10	0	0	0
8	200	200	120	100	100	40	10	0	0	0
9	200	200	120	100	100	22	10	0	0	0
10	200	200	120	100	100	44	10	0	0	0
11	200	200	120	100	100	52	10	10	0	0
12	200	200	120	100	100	50	10	10	10	0
13	200	200	120	100	100	48	10	10	10	0
14	200	200	120	100	100	48	10	10	0	0
15	200	200	120	100	100	35	10	10	0	0
16	200	200	120	100	100	42	10	0	0	0
17	200	200	120	100	100	40	10	0	0	0
18	200	200	120	100	100	55	10	10	0	0
19	200	200	120	100	100	57	10	10	10	0
20	200	200	120	100	100	53	10	10	0	0
21	200	200	120	100	100	37	10	10	0	0
22	200	200	120	100	100	21	10	0	0	0
23	200	200	120	100	100	42	10	0	0	0
24	200	200	120	100	96	10	10	0	0	0

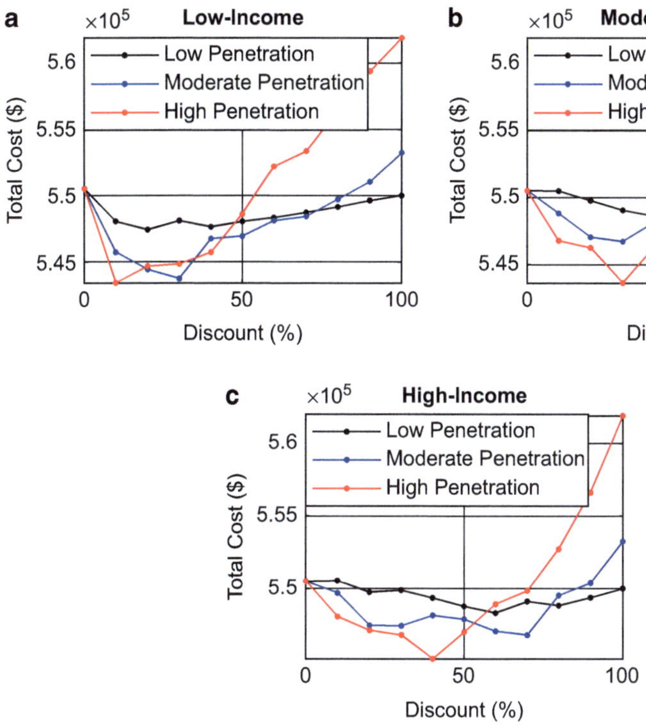

Fig. 2.10 Sensitivity analysis for the total cost of problem with respect to the value of incentive considering different PEV penetration levels for the (**a**) low-income, (**b**) moderate-income, and (**c**) high-income drivers

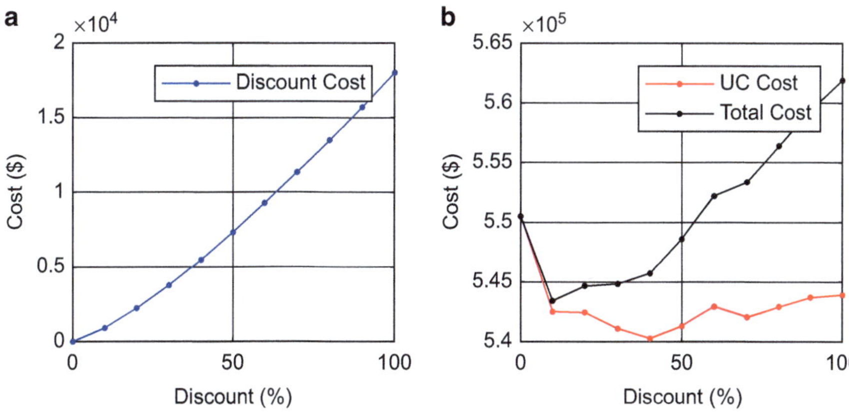

Fig. 2.11 Sensitivity analysis for (**a**) discount cost ($) and (**b**) UC cost and total cost ($) of low-income drivers with respect to the value of incentive considering the high PEV penetration level

2.5.3 *Effects of Unrealistic Modelling of Drivers' Social Class*

In this section, the effects of unrealistic modelling of drivers' social class on the minimum operation cost of system, optimal incentive, and generation scheduling and commitment of generation units are investigated considering different PEV penetration levels.

2.5.3.1 Minimum Operation Cost and Optimal Incentive

Figure 2.12a and c show the value of error in the minimum operation cost of system and optimal value of incentive due to the unrealistic modelling of drivers' social class (considering moderate-income instead of low-income) for different PEV penetration levels. In addition, Fig. 2.12b, d illustrate the value of error in the abovementioned parameters due to considering moderate-income instead of high-income social class for different PEV penetration levels. As can be seen, there are remarkable errors in the minimum operation cost of system and optimal value of incentive for each social class of drivers and any PEV penetration level.

2.5.3.2 Generation Scheduling and Commitment of Units

The values of error in the generation scheduling and commitment of generation units (G5 and G6), due to the unrealistic modelling of drivers' social class (moderate-income instead of low-income), are shown in Fig. 2.13 considering the high PEV penetration level. As can be seen, there are some errors in the generation level and status of units at some hours of the day.

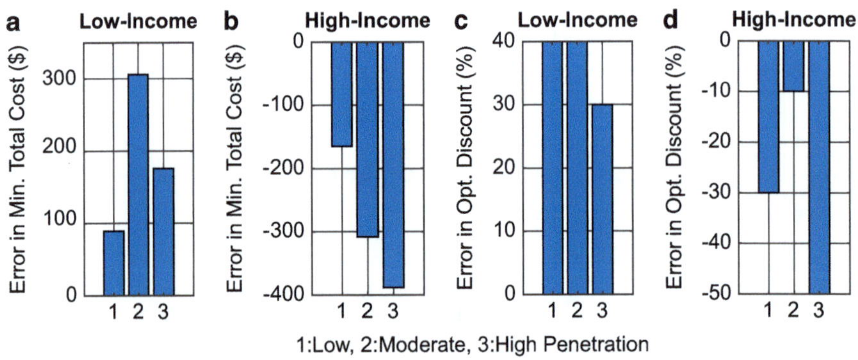

Fig. 2.12 Effects of unrealistic modelling of drivers' social class considering different PEV penetration levels. Error in minimum operation cost due to considering (**a**) moderate-income instead of low-income, (**b**) moderate-income instead of high-income. Error in optimal value of discount due to considering (**c**) moderate-income instead of low-income, (**d**) moderate-income instead of high-income

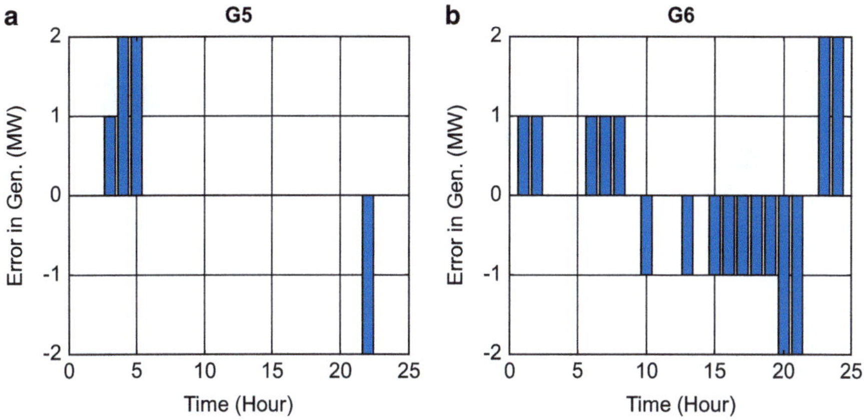

Fig. 2.13 Error in generation scheduling and commitment of generation units due to the unrealistic modelling of drivers' social class (moderate-income instead of low-income) considering the high PEV penetration level. (**a**) G5, (**b**) G6

Moreover, Fig. 2.14 illustrates the errors in the abovementioned parameters for the generation units G5–G9, due to considering moderate-income instead of high-income social class considering the high PEV penetration level. As can be seen, the values of error are larger.

2.6 Conclusion

The impacts of optimal fleet management (FM) of plug-in electric vehicles (PEVs) on the generation scheduling and commitment status of generation units were studied considering the social and technical aspects of problem. It was demonstrated that, by optimal FM of PEV fleet, although the FM imposes extra cost to the generation company (GENCO), the total operation cost of problem decreases for any PEV penetration level. Moreover, high PEV penetration level has more potential for the operation cost minimization.

By introducing the optimal incentive to the drivers, the responsive drivers shifted part of their charging demand from the peak period to the off-peak one, the system demand profile became flatter, the most expensive and pollutant generation units were shut down or their generation level decreased during the peak period, and the commitment and generation level of the least expensive and the least pollutant units increased during the off-peak period.

It was seen that, to have an optimal FM, a certain amount of incentive is needed for each social class of drivers and for each PEV penetration level. In addition, the FM of high-income drivers needs more incentive while it brings about the least cost

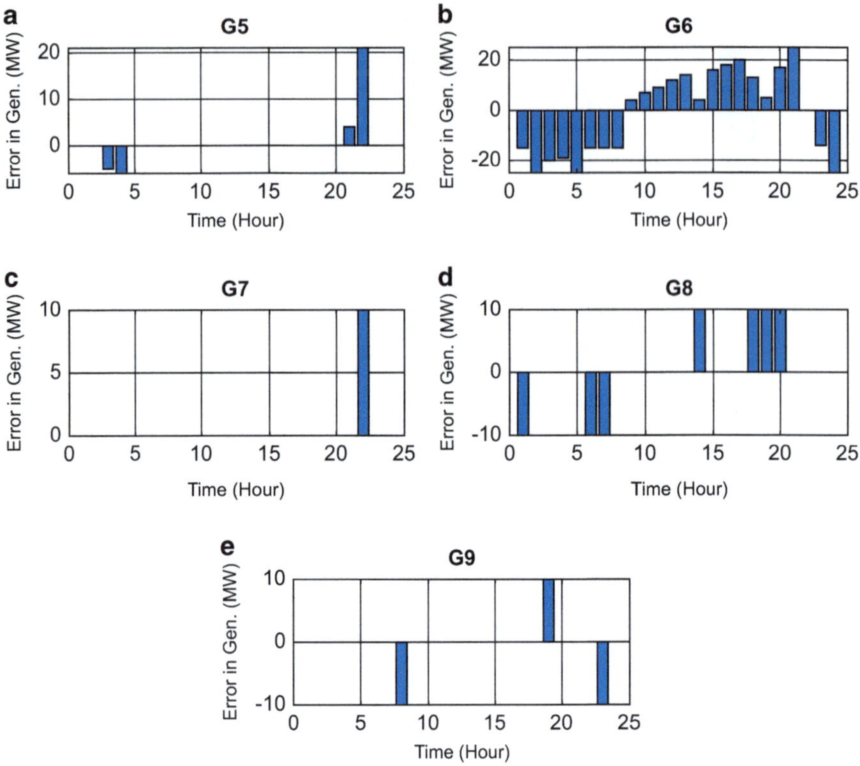

Fig. 2.14 Error in generation scheduling and commitment of generation units due to the unrealistic modelling of drivers' social class (moderate-income instead of high-income) considering the high PEV penetration level. (**a**) G5, (**b**) G6, (**c**) G7, (**d**) G8, (**e**) G9

saving for the GENCO. In other words, the FM of low-income drivers and high-income drivers has higher and lower potentials to minimize the operation cost of GENCO, respectively.

It was proven that the type of social class of drivers and even the PEV penetration level can change the optimal generation scheduling and commitment status of generation units. The unrealistic modelling of drivers' social class was investigated, and it was shown that remarkable errors occur in the minimum operation cost of system, optimal value of incentive, and generation scheduling and commitment of generation units.

As the general consequence, it can be concluded that the low-income and high-income drivers are the most and the least favorable drivers for GENCO from economic point of view, since they have the most and the least cooperation with the GENCO, respectively.

Nomenclature

Problem parameters	
β	Value of incentive (discount on charging fee of PEVs) in percent
ξ	Responsiveness probability of drivers (%)
μ	Mean of hourly Gaussian distribution function
σ	Standard deviation of hourly Gaussian distribution function
π^E	Electricity price ($/MWh)
$\alpha_1^E, \alpha_2^E, \alpha_3^E$	Emission coefficients of a generation unit in ton/MWh^2, ton/MW, and ton, respectively
$\alpha_1^F, \alpha_2^F, \alpha_3^F$	Fuel cost coefficients of a generation unit in $/MWh^2, $/MWh, and $, respectively
γ^E	Emission penalty factor in $/ton
C^{SHD}	Shut-down cost of a generation unit ($)
C^{STU}	Start-up cost of a generation unit ($)
$Cost^{INC}$	Cost of discount on charging fee of PEVs ($)
$Cost^F$	Fuel cost of generation units ($)
$Cost^E$	Greenhouse gas emissions cost of generation units ($)
$Cost^{SHD}$	Shut-down cost of a committed unit ($)
$Cost^{STU}$	Start-up cost of a decommitted unit ($)
D^{LOADs}	Loads demand (excluding PEVs demand) in MW
$\tilde{D}^{PEVs}$	Charging demand of responsive PEV fleet (MW)
$D^{PEVs,\ TOTAL}$	Total demand of PEVs (including responsive and nonresponsive PEVs) in MW
D^{PEVs}	Charging demand of nonresponsive PEVs (MW)
D^{TOTAL}	Total demand of system (MW)
e	Index of PEV
g, N^G	Index of generation unit and total number of generation units of GENCO
LI, MI, HI	Indices of low-income, moderate-income, and high-income drivers
MDT^G	Minimum "down" time limit of a generation unit (hour)
MUT^G	Minimum "up" time limit of a generation unit (hour)
$\tilde{N}^{PEVs}$	Expected number of responsive drivers
$N^{PEVs,\ TOTAL}$	Total number of PEVs in the area
OF	Objective function of problem ($)
$OFFT^G$	Time interval that the generation unit has been "off" (hour)
ONT^G	Time interval that the generation unit has been "on" (hour)
P^G	Power of a generation unit (MW)
$P^{G,\ MIN}$	Minimum power limit of a generation unit (MW)
$P^{G,\ MAX}$	Maximum power limit of a generation unit (MW)
P^{PEV}	Rated power of a PEV (kW)
RDR^G	Ramp-down rate limit of a generation unit (MW/hour)
RUR^G	Ramp-up rate limit of a generation unit (MW/hour)
s^G	Status of a generation unit, where "1" and "0" mean "on" and "off," respectively
SC	Social class of drivers
SOC^{PEVs}	State of charge of PEVs (%)
SR^{MIN}	Minimum required spinning reserve capacity of system (MW)
t	Index of time (hour)

(continued)

GA parameters	
θ^M	Mutation probability of a gene
θ^{PFBS}	Selection probability of a chromosome
a	Acceptance indicator of a chromosome for the new population
c	Index of a chromosome
f	Fitness of a chromosome
N^C	Number of chromosomes of population
r	A random number between [0,100] generated for each chromosome

Abbreviations

DR	Demand response
DSM	Demand side management
ERCOT	Electric Reliability Council of Texas
FERC	Federal Energy Regulatory Commission
FM	Fleet management
GA	Genetic algorithm
GENCO	Generation company
LI, MI, HI	Low-income, moderate-income, high-income
MPC	Model predictive control
PEV	Plug-in electric vehicle
PFBS	Probabilistic fitness-based selection
RES	Renewable energy source
SOC	State of charge
SCUC	Security-constrained unit commitment
UC	Unit commitment

References

1. "Monthly plug-in sales scorecard," Accessed on: Aug 2018. [Online]. Available: http://insideevs.com/monthly-plug-in-sales-scorecard
2. Clean Energy Ministerial, "Electric vehicles initiative," Accessed on: Aug. 2018. [Online]. Available: http://cleanenergyministerial.org/Our-Work/Initiatives/Electric-Vehicles. Accessed on: Aug 2018
3. [Online]. Available: https://about.bnef.com/blog/bps-energy-outlook-and-the-rising-consensus-on-ev-adoption/. Accessed on: Aug 2018
4. [Online]. Available: https://www.greentechmedia.com/articles/read/everyone-is-revising-electric-vehicle-forecasts-upward#gs.MbQ17ILt. Accessed on: Aug 2018
5. [Online]. Available: https://electrek.co/2017/05/05/electric-vehicle-sales-vs-gas-2040/. Accessed on: Aug 2018
6. Worldwide survey of network-driven demand-side management projects, Task XV, IEA-DSM Res. Rep. No. 1, 2006
7. IEA, "Strategic plan for the IEA demand-side management program 2008–2012" [Online]. Available: http://www.ieadsm.org

8. U.S. Department of Energy, Benefit of demand response in electricity markets and recommendations for achieving them, in: A report to the United States Congress, 2006 [Online]. Available: http://energy.gov/sites/prod/files/oeprod/DocumentsandMedia/DOEBenefitsofDemandResponseinElectricityMarketsandRecommendationsforAchievingThemReporttoCongress.pdf
9. M. Rahmani-Andebili, A. Abdollahi, M.P. Moghaddam, An investigation of implementing emergency demand response programs in unit commitment problem, in *IEEE Power & Energy Society General Meeting*, San Diego, pp. 1–7, 24–29 July 2011
10. M. Rahmani-Andebili, Investigating effect of responsive loads models on UC collaborated with demand side resources. IET Gener. Transm. Distrib. J. **7**(4), 420–430 (2013)
11. M. Rahmani-Andebili, Nonlinear demand response programs for residential customers with nonlinear behavioral models. Energ Buildings (Elsevier) **119**, 352–362 (2016)
12. N. Zhang, Z. Hu, D. Dai, S. Dang, M. Yao, Y. Zhou, Unit commitment model in smart grid environment considering carbon emissions trading. IEEE Trans. Smart Grid **7**(1), 420–427 (2016)
13. M. Rahmani-Andebili, G.K. Venayagamoorthy, Stochastic optimization for combined economic and emission dispatch with renewables, in *IEEE Symposium Series on Computational Intelligence*, Cape Town, pp. 1252–1258, 7–10 Dec 2015
14. M. Rahmani-Andebili, G.K. Venayagamoorthy, Combined emission and economic dispatch incorporating demand side resources, in *IEEE Clemson University Power System Conference*, Clemson, pp. 1–6, 10–13 Mar 2015
15. H. Ye, Z. Li, Robust security-constrained unit commitment and dispatch with recourse cost requirement. IEEE Trans. Power Syst. **31**(5), 3527–3536 (2016)
16. D.A. Tejada-Arango, P. Sanchez-Martın, A. Ramos, Security constrained unit commitment using line outage distribution factors. IEEE Trans. Power Syst. **33**(1), 329–337 (2018)
17. C. Zhao, R. Jiang, Distributionally robust contingency-constrained unit commitment. IEEE Trans. Power Syst. **33**(1), 94–102 (2018)
18. M. Rahmani-Andebili, Risk-cost based generation scheduling smartly mixed with reliability and market-driven demand response measures. Int. Trans. Electr. Energy Syst. **25**, 994–1007 (2015)
19. J. Meus, K. Poncelet, E. Delarue, Applicability of a clustered unit commitment model in power system modeling. IEEE Trans. Power Syst. **33**(2), 2195–2204 (2018)
20. S.I. Vagropoulos, G.A. Balaskas, A.G. Bakirtzis, An investigation of plug-in electric vehicle charging impact on power systems scheduling and energy costs. IEEE Trans. Power Syst. **32**(3), 1902–1912 (2017)
21. M. Rahmani-Andebili, G.K. Venayagamoorthy, Chapter III.VI: Co-operative responsive electric vehicles for social-economic dispatch, in *Cyber-Physical-Social Systems and Constructs in Electric Power Engineering*, (Stevenage, U.K., IET, 2016)
22. M. Rahmani-Andebili, Planning and operation of parking lots considering system, traffic, and drivers behavioral model. IEEE Trans. Syst. Man Cybern. Syst. (2018). https://doi.org/10.1109/TSMC.2018.2824122
23. M. Rahmani-Andebili, M. Fotuhi Firuzabad, An adaptive approach for PEVs charging management and reconfiguration of electrical distribution system penetrated by renewables. IEEE Trans. Ind. Inf. **14**(5), 2001 (2018)
24. D. Newbery, The economics of electric vehicles, in *EPRG and Imperial College London, E&E Seminar*, Cambridge, U.K., Jan 2013. [Online]. Available: http://www.eprg.group.cam.ac.uk/wpcontent/uploads/2013/01/EEJan13_EconomicsEVs.pdf
25. H. Saadat, *Power System Analysis* (McGraw-Hill, New York, 2009)
26. U.S. Energy Information Administration (EIA). [Online]. Available: http://www.eia.gov/todayinenergy/detail.cfm?id=9310. Accessed on Aug 2018
27. [Online]. Available: http://www.ferc.gov/market-oversight/mkt-electric/overview.asp. Accessed on Aug 2018

Chapter 3
Spinning Reserve Capacity Provision by the Optimal Fleet Management of Plug-In Electric Vehicles Considering the Technical and Social Aspects

3.1 Introduction

The electrification of transportation is seen as one of the solutions to the challenges including the global warming, the greenhouse gas emission, and the geopolitical concerns on availability of oil [1]. Since the plug-in electric vehicles (PEVs) are charged by the clean and free sources of energy, that is, renewable energy sources (RES) [2], they can mitigate the abovementioned issues [3, 4].

PEVs are being rolled out to the customers throughout the world and the USA. General Motors Company is now producing the Chevrolet Volt which is a plug-in hybrid electric vehicle (PHEV). Ford Motor and Nissan Motors Companies are manufacturing PEVs, that is, Ford Electric Focus and Leaf, respectively. Moreover, Tesla Motors is the producer of all-electric Tesla Roadster [5]. Figure 3.1 illustrates the projected annual and cumulative electric vehicle sales by 2040 [6]. As can be seen, there is a remarkable growth, where the electric vehicles will make up about 35% of all new vehicle sales in 2040.

As the battery prices fall, the unsubsidized electric vehicles are made as affordable as their gasoline counterparts. In other words, as the battery costs continue to fall, the demand for electric vehicles increases, since a battery cost accounts for almost one-third of total cost of an electric vehicle [6, 7]. According to a new analysis, regarding the electric vehicle market done by Bloomberg New Energy Finance (BNEF), the long-range electric vehicles will cost less than $22,000 (in today's dollars) by 2040 [6]. Moreover, the electric vehicles will cost the same as their internal combustion counterparts by 2022 that a lift-off point will occur for the electric vehicle sales [6].

Figure 3.2 shows the actual and estimated range of cost for the lithium-ion battery packs ($/kWh) and the yearly electricity demand of electric vehicles (GWh) for the period of 2010–2030. The descending trend of lithium-ion battery cost is the promising news; however, the sudden growth of electricity demand of electric

M. Rahmani-Andebili, *Planning and Operation of Plug-In Electric Vehicles*,
https://doi.org/10.1007/978-3-030-18022-5_3

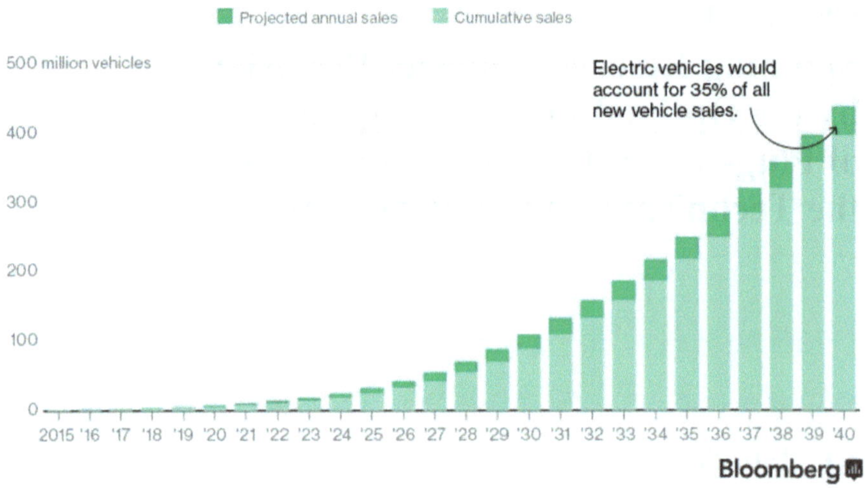

Fig. 3.1 The projected annual and cumulative electric vehicle sales by 2040 [6]. (Sources: Data compiled by Bloomberg New Energy Finance, Marklines)

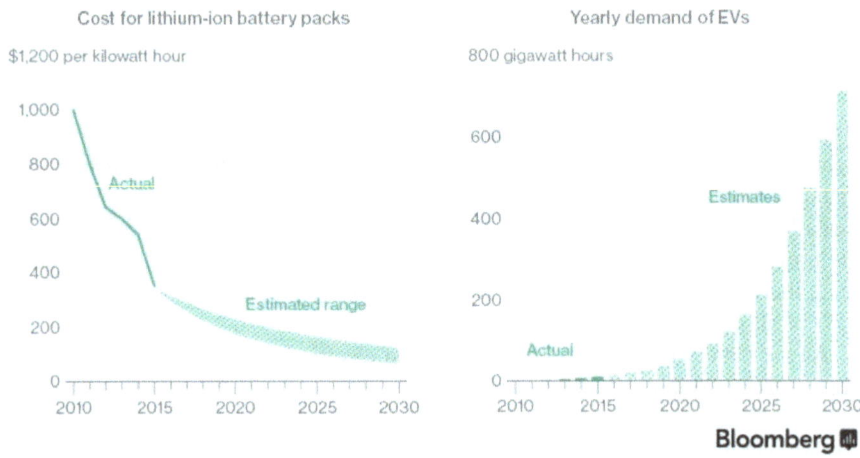

Fig. 3.2 The actual and estimated range of cost for the lithium-ion battery packs ($/kWh) and the yearly electricity demand of electric vehicles (GWh) for the period of 2010–2030 [6]. (Source: Data compiled by Bloomberg New Energy Finance)

vehicles is likely to be a treat for the electric power system that implies the necessity of having a strategy to deal with this phenomenon.

The rated power of a PEV's battery is very small compared to the capacity of an electric power system; however, if a fraction of internal combustion engine vehicles is replaced by the PEVs, a considerable capacity will be created as the energy storage. PEVs can act as the backup energy storage for the power system to provide vehicle-to-grid (V2G) service in the emergency faster than any generation unit. This feature of PEVs introduces them as a candidate to cooperate with the generation units to provide the spinning reserve capacity of power system.

In [8, 9], the participation of PEVs in the energy market has been studied. Reference [10] is the first study that has investigated the participation of PEVs in the spinning reserve market. In [11], the effect of changes in the spinning reserve market regulation on the PEVs and PEV aggregator has been studied. In [12], RESs and PEVs have been aggregated in a microgrid to participate in the energy and reserve markets. In [13], the participation of PEVs in the energy and reserve markets has been investigated. In [14], a stochastic approach has been applied to study the abovementioned problem. The main shortcoming of studies [10–14] is ignoring the social aspect of problem and the drivers' behavioral model in the spinning reserve capacity provision.

The contributions of chapter include:

- Modelling and studying the effects of optimal fleet management (FM) of PEVs on the spinning reserve capacity of power system.
- Classifying the drivers in three different social classes based on their reaction with respect to the value of incentive. Responsive drivers receive incentive for their availability at peak period to provide spinning reserve capacity in the normal condition. Moreover, they are offered discount on charging fee at off-peak period due to their V2G service provision in the contingency.
- Performing a sensitivity analysis about the system operation cost with respect to the value of incentive for each social class of drivers considering different PEV penetration levels.
- Studying the effects of unrealistic modelling of drivers' social class on the problem outcomes.

In the following, the social and technical aspects of problem are modelled and described in Sect. 3.2. In Sects. 3.3 and 3.4, the optimization approach and the problem formulation are presented, respectively. Section 3.5 deals with the problem simulation and results analysis, and finally Sect. 3.6 concludes the chapter.

3.2 Modelling the Social and Technical Aspects of Problem

3.2.1 Social Classification of Drivers

Due to the sociology and economic reasons, the drivers with different social classes do not react identically to the value of incentive suggested by a utility or an aggregator. Thus, the drivers' stratification seems to be unavoidable. In this chapter, the drivers are categorized in three different social classes based on their income level, namely, low-income, moderate-income, and high-income. Since the low-income and high-income drivers are more and less likely to be interested in the proposed incentive (β) to cooperate with the system operator, they can be considered as the most and the least responsive drivers, respectively.

In the behavior modelling of drivers' social classes, it is assumed that the responsiveness probability of a driver with any social class is zero if there is no incentive ($\beta = 0\%$) and it is 100% if the PEV is charged in free ($\beta = 100\%$). Regarding the incentives between 0% and 100%, a power function and an exponential function are applied to model the behavior of low-income and high-income drivers, since these functions return high and low values for the given small value of variable, respectively. Additionally, a linear function is applied to model the behavior of moderate-income drivers.

The responsiveness probability functions of low-income (LI), moderate-income (MI), and high-income (HI) drivers are presented in Eqs. (3.1, 3.2, and 3.3). Herein, t and e are the indices of time and PEV, respectively. Figure 3.3 shows the responsiveness probability of low-income, moderate-income, and high-income drivers with respect to the value of incentive.

The expected number of responsive drivers ($\tilde{N}^{PEVs}$) that will cooperate with the system operator is determined using Eq. (3.4). Herein, SC and $N^{PEVs,\ TOTAL}$ are the index of social class and the total number of PEVs in the area, respectively.

$$\xi_{e,LI,\beta,t} = 100 \times \left(\frac{\beta}{100}\right)^{0.3} \tag{3.1}$$

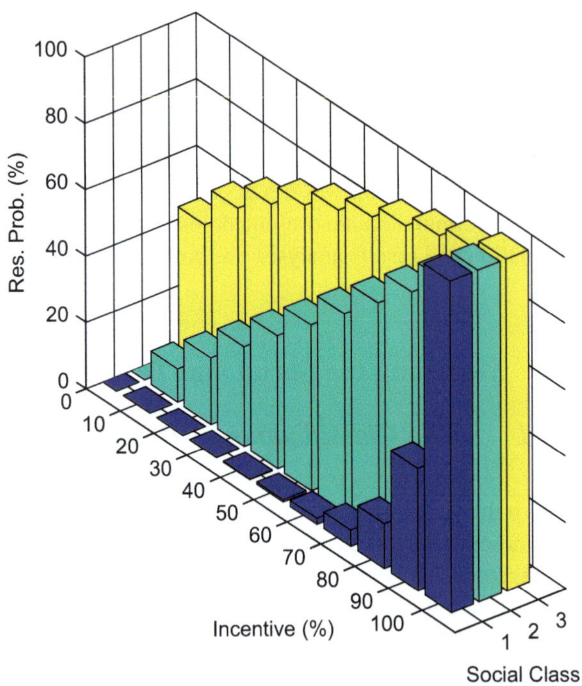

Fig. 3.3 Responsiveness probability of a driver with respect to the value of incentive ($\beta \in \{0\%, 10\%, \ldots, 100\%\}$) for each social class of driver (1: low-income, 2: moderate-income, 3: high-income)

$$\xi_{e,HI,\beta,t} = 100 \times exp\left(10 \times \left(\frac{\beta}{100} - 1\right)\right) \tag{3.2}$$

$$\xi_{e,MI,\beta,t} = 100 \times \left(\frac{\beta}{100}\right) \tag{3.3}$$

$$\tilde{N}^{PEVs}_{SC,\beta,t} = \sum_{e=1}^{N^{PEVs,TOTAL}} \frac{\xi_{e,SC,\beta,t}}{100}, \quad \forall SC \in \{LI, MI, HI\}, \quad \forall \beta \in \{0, 10, \ldots, 100\}, \quad \forall t \in \{1, \ldots, 24\} \tag{3.4}$$

3.2.2 Technical Aspect

Each responsive driver is offered an incentive for his/her availability at peak period to provide the spinning reserve capacity in the normal condition. The value of incentive in normal condition depends on the electricity price at peak period, the occurrence probability of contingency, the state of charge (SOC) and depth of discharge (DOD) of PEV, and the rated power of PEV.

In addition, another incentive, that is, the value of discount on charging fee at off-peak period, is considered for the responsive driver due to his/her V2G service provision in the contingency. The value of this incentive depends on the electricity price at off-peak period and the abovementioned parameters.

The hourly SOC of PEV fleet is modelled using the mean (μ) and standard deviation (σ) of hourly Gaussian distribution functions presented in [15]. Equation (3.5) presents the charging demand (MW) of responsive PEV fleet ($\tilde{D}^{PEVs}$) with social class of SC and incentive of β at time t. Herein, SOC^{PEVs}, P^{PEVs}, and $\tilde{N}^{PEVs}$ indicate the SOC of PEV fleet (%), the average rated power of battery of PEV fleet (kW), and the expected number of responsive drivers, respectively. In Eq. (3.6), $D^{PEVs,\ TOTAL}$ presents the total demand (MW) of PEV fleet (including responsive and nonresponsive PEVs), where $N^{PEVs,\ TOTAL}$ is the total number of PEVs in the area. Moreover, D^{PEVs} in Eq. (3.7) represents the hourly charging demand (MW) of nonresponsive PEV fleet. Demand of other loads (D^{LOADs}) is determined using Eq. (3.8), where D^{TOTAL} is the total demand of system (MW).

$$\tilde{D}^{PEVs}_{SC,\beta,t} = \left(1 - \frac{SOC^{PEVs}_t}{100}\right) \times \frac{P^{PEVs}}{1000} \times \tilde{N}^{PEVs}_{SC,\beta,t} \tag{3.5}$$

$$D^{PEVs,TOTAL}_t = \left(1 - \frac{SOC^{PEVs}_t}{100}\right) \times \frac{P^{PEVs}}{1000} \times N^{PEVs,TOTAL} \tag{3.6}$$

$$D_t^{PEVs} = D_t^{PEVs,TOTAL} - \tilde{D}_{SC,\beta,t}^{PEVs} \tag{3.7}$$

$$D_t^{LOADs} = D_t^{TOTAL} - D_t^{PEVs,TOTAL} \tag{3.8}$$

3.3 Optimization Technique

In this study, simulated annealing (SA) algorithm is applied as the optimization technique to solve the optimization problem. Herein, the daily operation cost of problem is defined as the value of internal energy of molten metal (ε^{SA}), and then it is tried to minimize the value of this energy.

The problem is solved for each possible value of incentive with a 10% step, that is, 0%, 10%, ..., 100%. Then, the optimal value of incentive is determined based on the minimum total cost of problem. In the following, the steps for applying SA in the optimization problem are presented and described.

- *Step 1: Obtaining the primary data*

  Parameters for applying SA: These parameters include the initial temperature of molten metal (θ_0^{SA}), the number of updates at each temperature (N^{SA}), and the value of coefficient to gradually decrease the temperature of molten metal (η^{SA}).

  Parameters of system: The value of all parameters of system and problem is obtained. Also, the value of incentive (β) is chosen.

  Updating the hourly spinning reserve capacities provided by the PEVs and generation units: The hourly capacity of spinning reserve provided by the responsive PEVs is updated based on the value of incentive, and then the hourly capacity of spinning reserve provided by the generation units is modified.

  Initial solution: A random solution for the problem variables is generated as an initial solution. The problem variables include the hourly status of generation units in the day presented in the SA matrix, as can be seen in Fig. 3.4.

- *Step 2: Generating an acceptable solution*

  Updating the solution: A random solution for the SA matrix is generated in the vicinity of old one.

  Checking the problem constraints: All the problem constraints are checked for the optimization time horizon, and if they are correct, the value of internal energy of molten metal is determined and the algorithm goes on; otherwise, the procedure is continued from Step 2. The value of internal energy of molten metal is determined by solving the optimal generation scheduling of generation units for the given SA matrix using the Lambda-Iteration Economic Dispatch method.

Fig. 3.4 The structure of SA matrix

	G1	G2	...
1	0/1	0/1	0/1
2	0/1	0/1	0/1
⋮	⋮	⋮	⋮
24	0/1	0/1	0/1

Checking the SA acceptance criterion: Based on the acceptance criterion (a^{SA}), presented in Eq. (3.9), the problem solution resulted in less internal energy of molten metal is always accepted; however, the solution with more value is accepted just by an adaptive probability (ρ^{SA}) presented in Eq. (3.10). Herein, r^{SA} is a random number in the range of [0,1). The value of adaptive probability decreases as the molten metal cools down.

$$a_k^{SA} = \begin{cases} 1 & \varepsilon_{k+1}^{SA} < \varepsilon_k^{SA} \; or \; \varepsilon_{k+1}^{SA} \geq \varepsilon_k^{SA} \& r_k^{SA} > \rho_k^{SA} \\ 0 & \varepsilon_{k+1}^{SA} \geq \varepsilon_k^{SA} \& r_k^{SA} \leq \rho_k^{SA} \end{cases} \tag{3.9}$$

$$\rho_k^{SA} = e^{-\frac{\varepsilon_{k+1}^{SA} - \varepsilon_k^{SA}}{\theta_k^{SA}}} \tag{3.10}$$

- *Step 3: Checking the number of iterations for the current temperature*
  If the number of iterations in the current temperature is not equal to the predefined value (N^{SA}), the process is continued form Step 2; otherwise, the temperature of molten metal is decreased based on Eq. (3.11).

$$\theta_{k+1}^{SA} = \eta^{SA} \times \theta_k^{SA} \tag{3.11}$$

- *Step 4: Concluding*

Checking the temperature of molten metal: The temperature of molten metal is measured and if the molten metal is frozen ($\theta^{SA} \approx 0$), the optimization procedure is terminated; otherwise, it is continued from Step 2.

Introducing the outcomes: The consequences include the minimum daily operation cost of problem and the optimal hourly value of spinning reserve capacity provided by the responsive PEVs and generation units.

3.4 Problem Formulation

3.4.1 Objective Function of Problem

The objective function of problem is presented in Eq. (3.12) that includes the incentive cost related to the normal condition ($Cost^{INC,\ NORMAL}$), the cost of discount on charging fee related to the contingency ($Cost^{INC,\ CONTINGENCY}$), the fuel cost of generation units ($Cost^F$), the greenhouse gas emission cost of generation units ($Cost^E$), the start-up cost of decommitted units ($Cost^{STU}$), and the shut-down cost of committed units ($Cost^{SHD}$). Herein, g and N^G are the index of generation unit and the total number of generation units, respectively.

$$OF = min\left\{\begin{array}{l}\sum\limits_{t\epsilon PEAK} Cost_{SC,\beta,t}^{INC,NORMAL} + \sum\limits_{t\epsilon OFFPEAK} Cost_{SC,\beta,t}^{INC,CONTINGENCY} \\ +\sum\limits_{t=1}^{24}\sum\limits_{g=1}^{N^G}\left[Cost_{g,t}^{F} + Cost_{g,t}^{E} + Cost_{g,t}^{STU} + Cost_{g,t}^{SHD}\right]\end{array}\right\} \tag{3.12}$$

3.4.2 Cost Terms of Problem

In the following, the cost terms of objective function are described. Herein, all the cost terms are in US dollar.

Incentive cost in normal condition As can be seen in Eq. (3.13), this cost term includes the incentive paid to the responsive drivers for their availability at peak period to provide spinning reserve capacity in the normal condition. In Eq. (3.13), π^E, $\tilde{SR}^{PEVs}$, $Prob^{CONTINGENCY}$, and X are the electricity price (\$/MWh), the spinning reserve capacity (MW) provided by the PEVs at peak period in normal condition, the occurrence probability of contingency (%), and a binary integer as the contingency indicator (if $X = 1$), respectively. The value of $\tilde{SR}^{PEVs}$ is determined using Eq. (3.14), where SOC^{PEVs}, DOD^{PEVs}, P^{PEVs}, and $\tilde{N}^{PEVs}$ are the SOC of responsive PEVs (%), the DOD defined for the battery of a PEV (%), the rated power of a PEV (kW), and the expected number of responsive drivers, respectively.

$$Cost_{SC,\beta,t}^{INC,NORMAL} = \pi_t^E \times \frac{\beta}{100} \times \tilde{SR}_t^{PEVs} \times \left(1 - \frac{Prob^{CONTINGENCY}}{100}\right) \times (1 - X) \tag{3.13}$$

$$\tilde{SR}_t^{PEVs} = \frac{SOC_t^{PEVs} - DOD^{PEVs}}{100} \times \frac{P^{PEVs}}{1000} \times \tilde{N}_{SC,\beta,t}^{PEVs} \tag{3.14}$$

Incentive cost in contingency This cost term includes the cost of discount on charging fee of PEVs at off-peak period offered to the responsive drivers due to their V2G service provision in the contingency at peak period. In Eq. (3.15), $\tilde{V2G}^{PEVs}$ is the V2G power (MW) provided by the PEVs in the contingency, which is equal to $\tilde{SR}_t^{PEVs}$, as can be seen in Eq. (3.16).

$$Cost_{SC,\beta,t}^{INC,CONTINGENCY} = \pi_t^E \times \left(1 + \frac{\beta}{100}\right) \times \tilde{V2G}_t^{PEVs} \times \frac{Prob^{CONTINGENCY}}{100} \times X \tag{3.15}$$

$$\tilde{V2G}_t^{PEVs} = \tilde{SR}_t^{PEVs} \tag{3.16}$$

Fuel cost of generation units The fuel cost of each generation unit ($Cost^F$) is considered a quadratic polynomial of its power (P^G) in MW [16]. The α_1^F, α_2^F, and α_3^F are the fuel cost coefficients of a generation unit in $/MWh^2, $/MWh, and $, respectively.

$$Cost_{g,t}^F = \alpha_{1,g}^F \times \left(P_{g,t}^G\right)^2 + \alpha_{2,g}^F \times \left(P_{g,t}^G\right) + \alpha_{3,g}^F \tag{3.17}$$

Greenhouse gas emissions cost of generation units The greenhouse gas emissions cost of each generation unit is also considered a quadratic polynomial of its power (P^G) [16]. The α_1^E, α_2^E, and α_3^E are the emission coefficients of a generation unit in Ton/MWh^2, Ton/MW, and Ton, respectively. Moreover, γ^E is the emission penalty factor in $/Ton.

$$Cost_{g,t}^E = \gamma^E \times \left(\alpha_{1,g}^E \times \left(P_{g,t}^G\right)^2 + \alpha_{2,g}^E \times \left(P_{g,t}^G\right) + \alpha_{3,g}^E\right) \tag{3.18}$$

Start-up cost and shut-down cost of generation units The start-up cost of a decommitted unit ($Cost^{STU}$) and shut-down cost of a committed unit ($Cost^{SHD}$) are presented in Eqs. (3.19) and (3.20), respectively. Herein, s^G indicates the status of a generation unit, where "1" and "0" mean "on" and "off," respectively. Moreover, C^{STU} and C^{SHD} are the start-up and shut-down costs, respectively.

$$Cost_{g,t}^{STU} = C_g^{STU} \times \left(1 - s_{g,t-1}^G\right) \times s_{g,t}^G \tag{3.19}$$

$$Cost_{g,t}^{SHD} = C_g^{SHD} \times s_{g,t-1}^G \times \left(1 - s_{g,t}^G\right) \tag{3.20}$$

3.4.3 Constraints of Problem

System power balance constraint This constraint is presented in Eq. (3.21), where the responsive PEVs have electricity demand in the normal condition, but provide V2G power in the contingency.

$$\sum_{g=1}^{N^G} P_{g,t}^{G} \times s_{g,t}^{G} = D_t^{LOADs} + D_t^{PEVs} + \left[\tilde{D}_{SC,\beta,t}^{PEVs} \times (1-X) + \tilde{V2G}_t^{PEVs} \times X\right] \tag{3.21}$$

Minimum generation constraint The generation units that are in "on" status must be able to supply the minimum demand of system including the load demand, the demand of nonresponsive PEVs, and the demand of responsive PEVs in the normal condition or their V2G power provision in the contingency, as can be seen in Eq. (3.22).

$$\sum_{g=1}^{N^G} P_{g}^{G,MIN} \times s_{g,t}^{G} \leq D_t^{LOADs} + D_t^{PEVs} + \left[\tilde{D}_{SC,\beta,t}^{PEVs} \times (1-X) + \tilde{V2G}_t^{PEVs} \times X\right] \tag{3.22}$$

Maximum generation constraint considering spinning reserve capacity The generation units that are in "on" status must be able to supply the maximum demand of system including the load demand, the demand of nonresponsive PEVs, and the demand of responsive PEVs in the normal condition or their V2G power provision in the contingency, considering the minimum requirement of spinning reserve capacity of system (SR^{MIN}), as can be seen in Eq. (3.23). The SR^{MIN} includes the spinning reserve capacity provided by the responsive PEVs ($\tilde{SR}_t^{PEVs}$) and the generation units (SR_t^{Gs}), as can be seen in Eq. (3.24). SR^{Gs} is determined using Eq. (3.25).

$$\sum_{g=1}^{N^G} P_{g}^{G,MAX} \times s_{g,t}^{G} \geq D_t^{LOADs} + D_t^{PEVs} + \left[\tilde{D}_{SC,\beta,t}^{PEVs} \times (1-X) + \tilde{V2G}_t^{PEVs} \times \mathrm{X}\right] + SR_t^{MIN} \tag{3.23}$$

$$SR_t^{MIN} = SR_t^{Gs} + \tilde{SR}_t^{PEVs} \tag{3.24}$$

$$SR_t^{Gs} = \sum_{g=1}^{N^G} \left(P_g^{G,MAX} - P_g^G\right) \times s_{g,t}^G \tag{3.25}$$

Power constraints of generation units The maximum and minimum power constraints of a generation unit at every hour of operation period are presented in Eq. (3.26).

$$P_g^{G,MIN} \le P_{g,t}^G \le P_g^{G,MAX} \tag{3.26}$$

Ramp-up rate and ramp-down rate constraints of generation units The ramp-up rate (RUR^G) and ramp-down rate (RDR^G) constraints of a generation unit at every hour of operation period are presented in Eqs. (3.27) and (3.28), respectively.

$$\left(P_{g,t+1}^G - P_{g,t}^G\right) \le RUR_g^G \tag{3.27}$$

$$\left(P_{g,t}^G - P_{g,t+1}^G\right) \le RDR_g^G \tag{3.28}$$

Minimum "down" time and minimum "up" time constraints of generation units The minimum "down" time (MDT^G) and minimum "up" time (MUT^G) constraints of a generation unit at every hour of operation period are presented in Eqs. (3.29) and (3.30), respectively. Herein, $OFFT^G$ and ONT^G indicate the time intervals that the generation unit has been "off" and "on," respectively.

$$OFFT_{g,t}^G \ge MDT_g^G \tag{3.29}$$

$$ONT_{g,t}^G \ge MUT_g^G \tag{3.30}$$

3.5 Simulation Results

3.5.1 Base Case

Table 3.1 presents the number of PEVs in the low, moderate, and high PEV penetration levels. The hourly demand of PEV fleet is calculated using their hourly SOC level (presented in [15]) and assuming Nissan Leaf as the type of PEVs with the rated power and capacity of 24 kW and 24 kWh, respectively [17]. In this study, DOD is considered about 20%.

Table 3.1 Number of PEVs in each penetration level

Penetration level	Low	Moderate	High
Number of PEVs	1000	3000	5000

Table 3.2 The technical characteristics of generation units

Generation unit	**G1**	**G2**	**G3**	**G4**	**G5**
α_1^F (/MWh^2)	0.00048	0.00031	0.00200	0.00211	0.00398
α_2^F (/MWh)	16.19	17.26	16.60	16.50	19.70
α_3^F ($)	1000	970	700	680	450
α_1^E (Ton/MWh^2)	0.0005	0.0005	0.0005	0.0005	0.0010
α_2^E (Ton/MWh)	0.4050	0.4320	0.4150	0.4120	0.4930
α_3^E (Ton)	0.3000	0.4250	0.4500	0.7000	0.7250
P^{MIN} (MW)	75	75	15	15	15
P^{MAX} (MW)	200	200	120	100	100
MUT (h)	5	5	5	5	5
MDT (h)	−5	−5	−5	−5	−5
RUR (MW/h)	110	110	80	80	80
RDR (MW/h)	110	110	80	80	80
C^{STU} ($)	4500	5000	550	560	900
C^{SHD} ($)	4500	5000	550	560	900
Initial status	+24	+24	+24	+24	+24
Generation unit	**G6**	**G7**	**G8**	**G9**	**G10**
α_1^F (/MWh^2)	0.00712	0.00790	0.00813	0.00822	0.00873
α_2^F (/MWh)	22.26	27.74	25.92	27.27	27.79
α_3^F ($)	370	480	660	665	670
α_1^E (Ton/MWh^2)	0.0020	0.0020	0.0024	0.0025	0.0025
α_2^E (Ton/MWh)	0.5560	1.0940	1.6480	1.6820	1.6950
α_3^E (Ton)	0.9250	1.2000	1.6500	1.6625	1.7750
P^{MIN} (MW)	10	10	10	10	10
P^{MAX} (MW)	80	50	25	20	20
MUT (h)	3	1	1	1	1
MDT (h)	−3	−1	−1	−1	−1
RUR (MW/h)	60	20	10	10	10
RDR (MW/h)	60	20	10	10	10
C^{STU} ($)	170	30	30	30	30
C^{SHD} ($)	170	30	30	30	30
Initial status	−3	−1	−1	−1	−1

The power generation system is a 10-unit system with the technical specifications presented in Table 3.2. The minimum required spinning reserve capacity of system is about 10% of total demand at each hour. The occurrence probability of a contingency is assumed about 10%. The amount of penalty for greenhouse gas emissions is

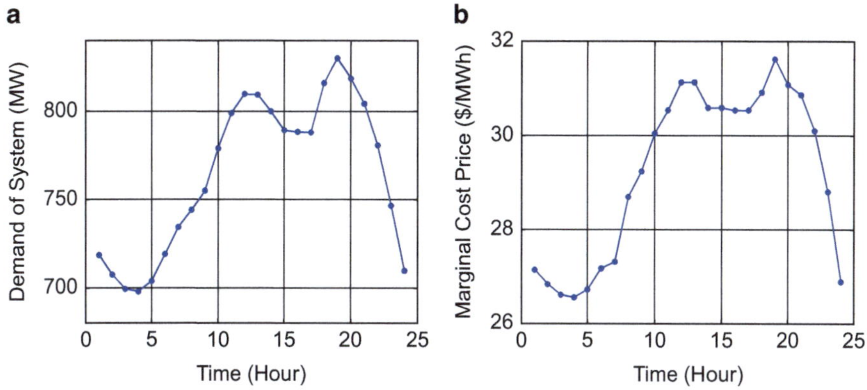

Fig. 3.5 (**a**) System demand profile (MW) [19]. (**b**) Hourly marginal cost price ($/MWh)

about $10/Ton based on the California Air Resources Board auction of greenhouse gas emissions [18]. Figure 3.5a shows the hourly demand profile of system (MW) that its per unit value is the same as the demand profile of Electric Reliability Council of Texas (ERCOT) power market [19]. Herein, the demand profile above 0.9 p.u. is considered as the peak period (Hours: 9–22). In addition, Fig. 3.5b illustrates the hourly marginal cost price ($/MWh) calculated based on the results of generation scheduling and unit commitment (UC) problem of system before FM. The hourly marginal cost price is applied as the hourly electricity price.

3.5.1.1 Low-Income Drivers

The problem simulation results before and after optimal FM of low-income drivers considering the low, moderate, and high PEV penetration levels are presented in Tables 3.3, 3.4, 3.5, and 3.6 and Figs. 3.6, 3.7, and 3.8. As can be seen in Table 3.3, the system operation cost is about 0.55050 million $/day before FM of PEVs. However, after optimal FM and the cooperation of PEVs with generation units to provide spinning reserve capacity, there are about $3109/day, $6895/day, and $10305/day cost savings for the low, moderate, and high PEV penetration levels, respectively. As can be seen, the incentive cost in normal condition makes up most of the total incentive cost. Moreover, as can be noticed, the value of optimal incentive needs to be investigated and updated for each penetration level.

Figures 3.6b and 3.7b illustrate the optimal hourly spinning reserve capacity in normal condition provided by the generation units and PEVs after optimal FM of PEVs for the low penetration (with 10% incentive) and high penetration (with 30% incentive), respectively. Moreover, Figs. 3.6c and 3.7c demonstrate the value of difference in the hourly spinning reserve capacity of system before and after FM of PEVs for the low and high PEV penetrations, respectively. In addition, the daily spinning reserve capacity percentage provided by the generation units and PEVs in

Table 3.3 Problem simulation results before and after optimal FM of low-income drivers considering different PEV penetration levels

Outputs	Before FM	After optimal FM		
	Any penetration	Low penetration	Moderate penetration	High penetration
Optimal incentive (%)	0	10	30	30
Incentive cost in normal condition ($)	0	169	2121	3535
Incentive cost in contingency ($)	0	2	23	39
Minimum total cost (Million $/day)	0.55050	0.54739	0.54360	0.54019
Cost saving ($/day)	–	3109	6895	10305

Table 3.4 Commitment status of generation units in the normal condition before FM

Generation unit	Time (hour)
G1	111111111111111111111111
G2	111111111111111111111111
G3	111111111111111111111111
G4	111111111111111111111111
G5	111111111111111111111111
G6	111111111111111111111111
G7	000000111111111111111110
G8	000000000111111111111100
G9	000000000011110001111000
G10	000000000000000001110000

Table 3.5 Commitment status of generation units in the normal condition after optimal FM of low-income drivers considering the low PEV penetration level (10% incentive)

Generation unit	Time (Hour)
G8	000000000011111111111100
G9	000000000001100001111000
G10	000000000000000000110000

the normal condition after FM of PEVs for the low and high PEV penetrations are shown in Figs. 3.6d and 3.7d, respectively.

As can be seen in Fig. 3.6b, the PEVs provide a small part of spinning reserve capacity of system at peak period, but it causes remarkable updates in the hourly spinning reserve capacity of system, as can be seen in Fig. 3.6c. As can be seen in Fig. 3.7b and d, a noticeable portion of spinning reserve capacity of system is provided by the PEV fleet considering high PEV penetration level.

Table 3.6 Commitment status of generation units in the normal condition after optimal FM of low-income drivers considering the high PEV penetration level (30% incentive)

Generation unit	Time (Hour)
G7	000000110111111111111110
G8	000000000000000001111100
G9	000000000000000000110000
G10	000000000000000000000000

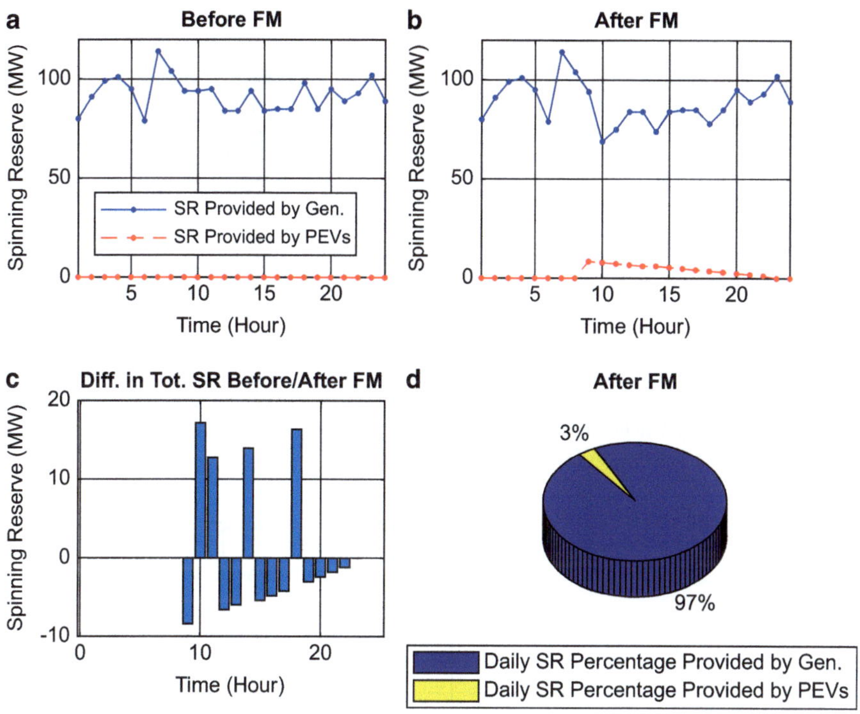

Fig. 3.6 Low-income drivers considering the low PEV penetration level. Optimal hourly spinning reserve (SR) capacity (MW) in normal condition provided by the generation units and PEVs (**a**) before and (**b**) after FM (10% incentive). (**c**) Value of difference (MW) in the hourly total spinning reserve capacity of system before/after FM. (**d**) Daily spinning reserve capacity percentage provided by the generation units and PEVs in the normal condition

The hourly value of V2G power and grid-to-vehicle (G2V) demand in the emergency considering different PEV penetration levels is shown in Fig. 3.8. As can be seen, the PEVs provide V2G service at peak period and receive the same

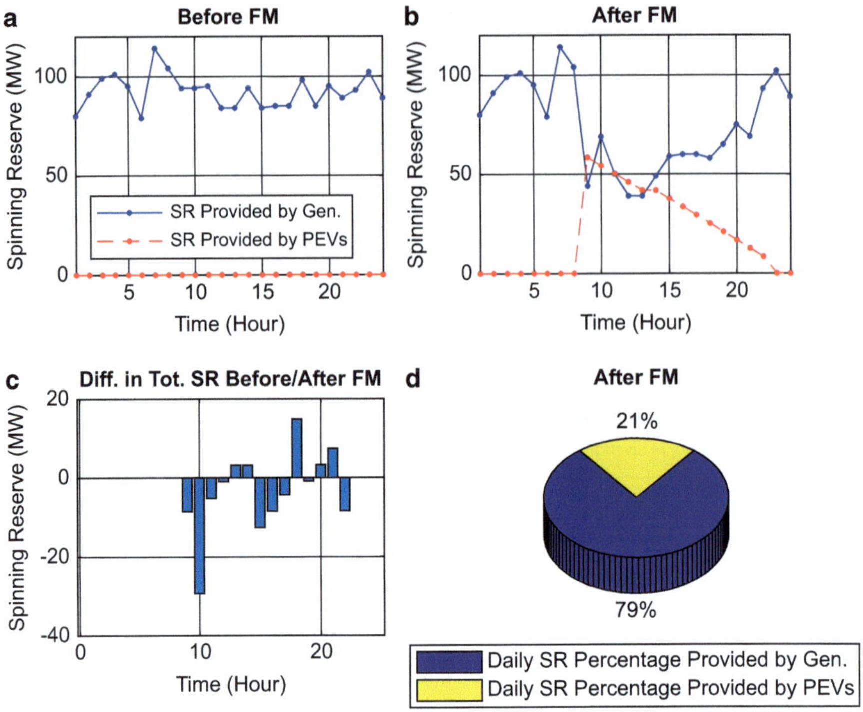

Fig. 3.7 Low-income drivers considering the high PEV penetration level. Optimal hourly spinning reserve (SR) capacity (MW) in the normal condition provided by the generation units and PEVs (**a**) before and (**b**) after FM (30% incentive). (**c**) Value of difference (MW) in the hourly total spinning reserve capacity of system before/after FM. (**d**) Daily spinning reserve capacity percentage provided by the generation units and PEVs in the normal condition

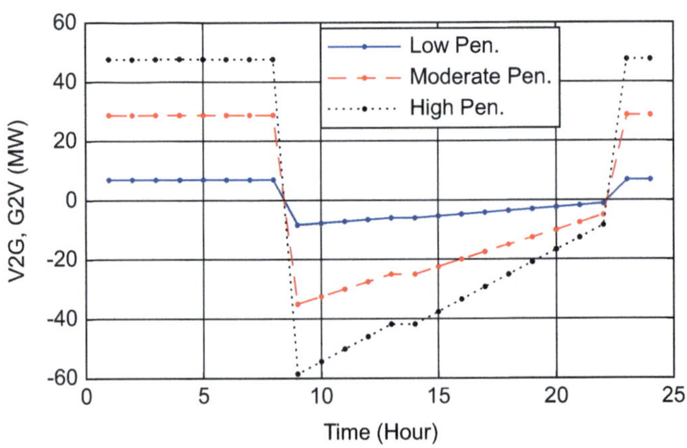

Fig. 3.8 Hourly value of G2V demand (MW) and V2G power provided by the low-income drivers after FM in the emergency considering different PEV penetration levels

amount of energy at off-peak period in the form of G2V service. Table 3.4 presents the commitment status of generation units in the normal circumstance before FM. In addition, Tables 3.5 and 3.6 show the commitment status of generation units in the normal condition after optimal FM of low-income drivers considering the low (10% incentive) and high (30% incentive) penetration levels, respectively. The statuses highlighted in red color indicate the differences compared to the results before FM. As can be seen, by optimal FM of PEVs, the commitment of most expensive and pollutant generation units decreases. This consequence is more extensive in the case of high PEV penetration level.

3.5.1.2 Moderate-Income Drivers

Table 3.7 presents the simulation results before and after optimal FM of moderate-income drivers considering different PEV penetration levels. As can be seen, compared to the results of FM of low-income drivers, the values of cost saving decrease for any PEV penetration level, while more incentive is paid to the moderate-income drivers.

3.5.1.3 High-Income Drivers

The problem simulation results before and after optimal FM of high-income drivers considering different PEV penetration levels are presented in Tables 3.8, 3.9, and 3.10 and Figs. 3.9, 3.10, and 3.11. By comparing Tables 3.8, 3.7, and 3.3, it is noticed that optimal FM of high-income drivers results in the least cost saving while they receive the most value of incentive.

In addition, by comparing Fig. 3.10b and d with Fig. 3.7b and d (or similarly Fig. 3.9b and d with Fig. 3.6b and d), it can be observed that less spinning reserve capacity is assigned to the high-income drivers. Moreover, the high-income drivers

Table 3.7 Problem simulation results before and after optimal FM of moderate-income drivers considering different PEV penetration levels

	Before FM	After optimal FM		
Outputs	Any penetration	Low penetration	Moderate penetration	High penetration
Optimal incentive (%)	0	50	70	50
Incentive cost in normal condition ($)	0	845	4971	4227
Incentive cost in contingency ($)	0	9	54	46
Minimum total cost (Million $/day)	0.55050	0.54807	0.54657	0.54499
Cost saving ($/day)	–	2428	3922	5509

Table 3.8 Problem simulation results before and after optimal FM of high-income drivers considering different PEV penetration levels

	Before FM	After optimal FM		
Outputs	Any penetration	Low penetration	Moderate penetration	High penetration
Optimal incentive (%)	0	90	80	90
Incentive cost in normal condition ($)	0	1120	1098	5597
Incentive cost in contingency ($)	0	12	12	62
Minimum total cost (Million $/day)	0.55050	0.54997	0.54979	0.54841
Cost saving ($/day)	–	530	710	2090

Table 3.9 Commitment status of generation units in normal condition after optimal FM of high-income drivers (90% incentive) considering the low PEV penetration level

Generation unit	Time (Hour)
G9	000000000001110001111000
G10	000000000000000000110000

Table 3.10 Commitment status of generation units in the normal condition after optimal FM of high-income drivers (90% incentive) considering the high PEV penetration level

Generation unit	Time (Hour)
G7	000000110111111111111110
G8	000000000011110011111100
G9	000000000000000001111000
G10	000000000000000000100000

have less G2V demand and V2G power in the emergency for any PEV penetration level, as can be noticed by comparing Fig. 3.8 with Fig. 3.11.

Also, Tables 3.9 and 3.10 present the commitment status of generation units in the normal condition after optimal FM of high-income drivers considering the low (90% incentive) and high (90% incentive) penetration levels, respectively. By comparing Table 3.10 with Table 3.6 (or similarly Tables 3.9 and 3.5), it can be concluded that fewer expensive and pollutant generation units are shut down by optimal FM of high-income drivers.

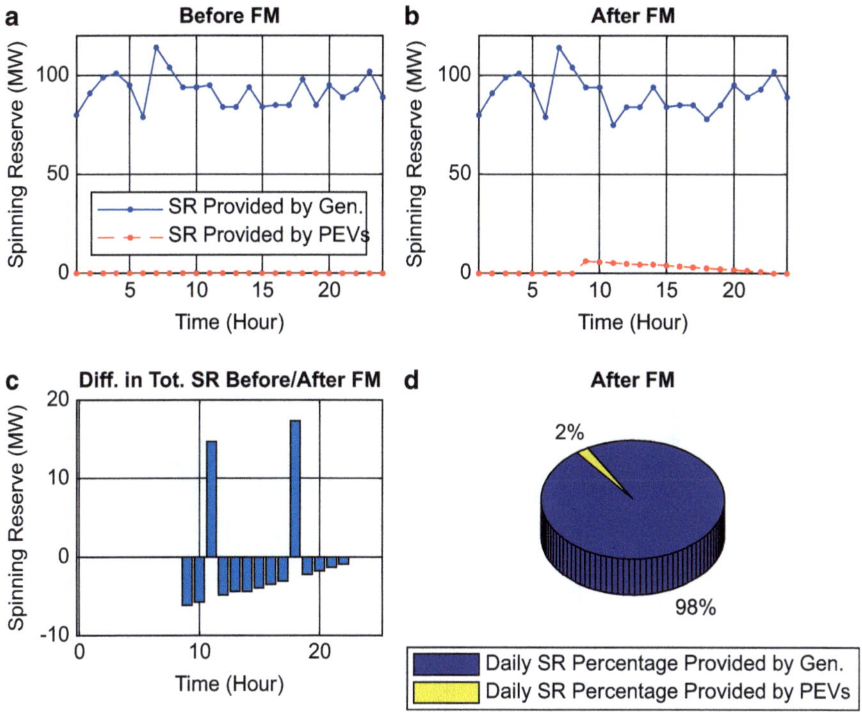

Fig. 3.9 High-income drivers considering the low PEV penetration level. Optimal hourly spinning reserve (SR) capacity (MW) in the normal condition provided by the generation units and PEVs (**a**) before and (**b**) after FM (90% incentive). (**c**) Value of difference (MW) in the hourly total spinning reserve capacity of system before/after FM. (**d**) Daily spinning reserve capacity percentage provided by the generation units and PEVs in the normal condition

3.5.2 Sensitivity Analysis for Total Cost of Problem

The sensitivity analysis for the minimum operation cost of problem with respect to the value of incentive for different PEV penetration levels and social classes of drivers is shown in Fig. 3.12. As can be seen, the sensitivity curves are nonlinear and distinct for each PEV penetration level, as well as for each social class of drivers. Therefore, it is necessary to find the optimal value of incentive for the responsive drivers to motivate them to participate in the spinning reserve capacity provision of system.

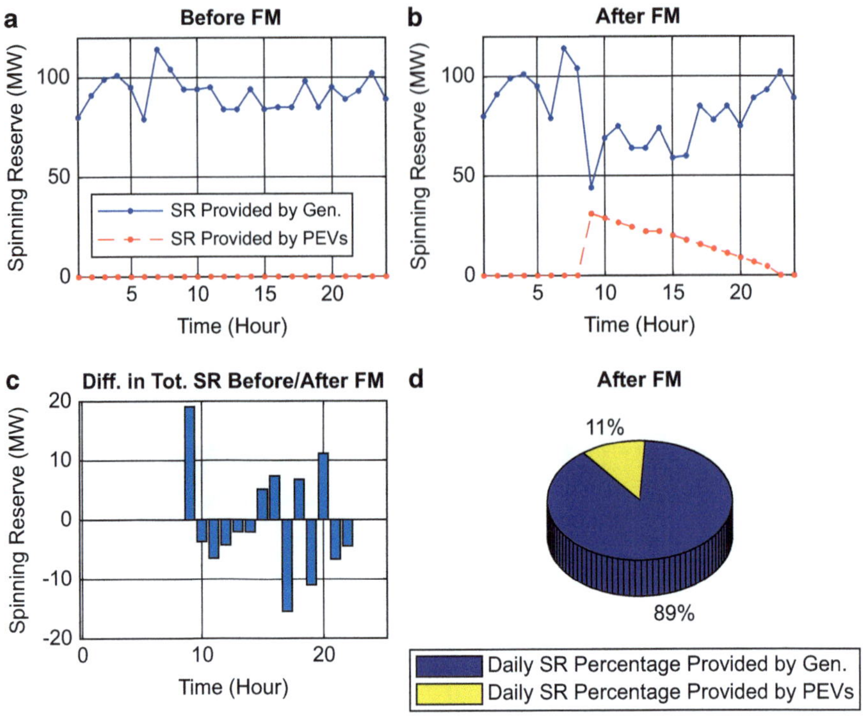

Fig. 3.10 High-income drivers considering the high PEV penetration level. Optimal hourly spinning reserve (SR) capacity (MW) in the normal condition provided by the generation units and PEVs (**a**) before and (**b**) after FM (90% incentive). (**c**) Value of difference (MW) in the hourly total spinning reserve capacity of system before/after FM. (**d**) Daily spinning reserve capacity percentage provided by the generation units and PEVs in the normal condition

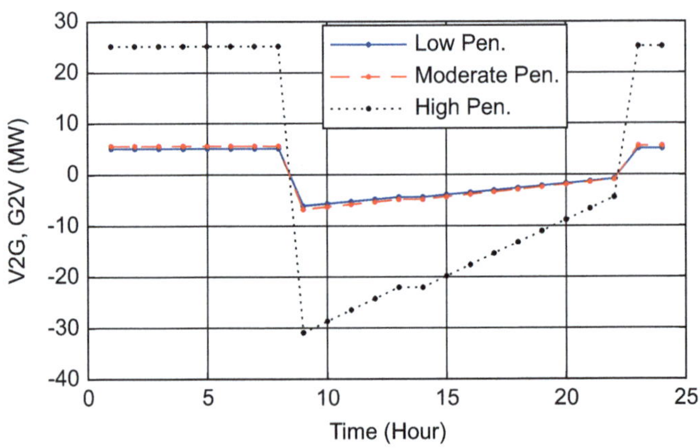

Fig. 3.11 Hourly value of G2V demand (MW) and V2G power provided by the high-income drivers after FM in the emergency considering different PEV penetration levels

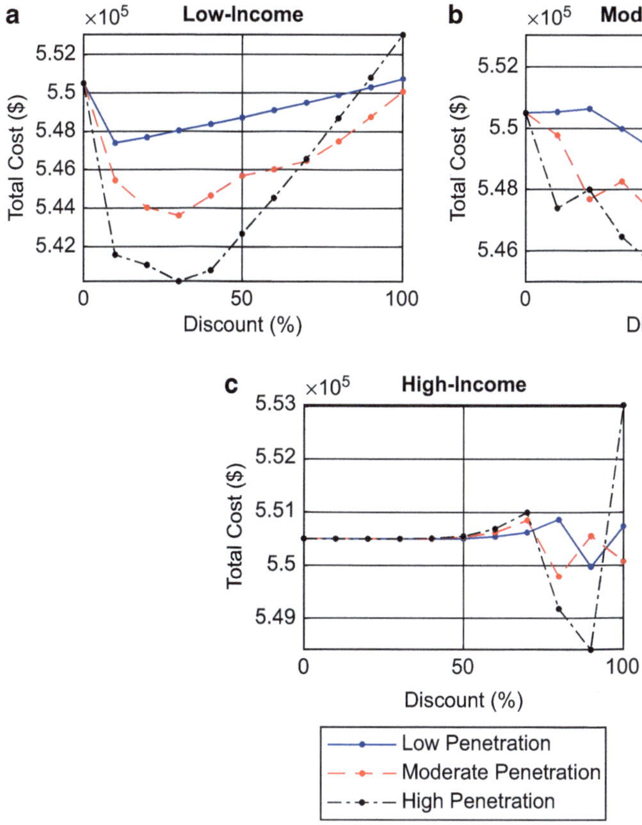

Fig. 3.12 Sensitivity analysis for the minimum total cost of problem ($) with respect to the value of incentive (%) for (**a**) low-income, (**b**) moderate-income, and (**c**) high-income drivers and for each PEV penetration level

3.5.3 *Effects of Unrealistic Modelling of Drivers' Social Class*

Figure 3.13 illustrates the effects of unrealistic modelling of drivers' social class on the minimum operation cost of problem and optimal value of incentive. In other words, Fig. 3.13a and c demonstrate the values of error in the minimum operation cost of problem and optimal value of incentive, due to the consideration of moderate-income instead of low-income social class for the drivers. Moreover, Fig. 3.13b, d show the values of error in the abovementioned parameters caused by the consideration of moderate-income instead of high-income social class for the drivers.

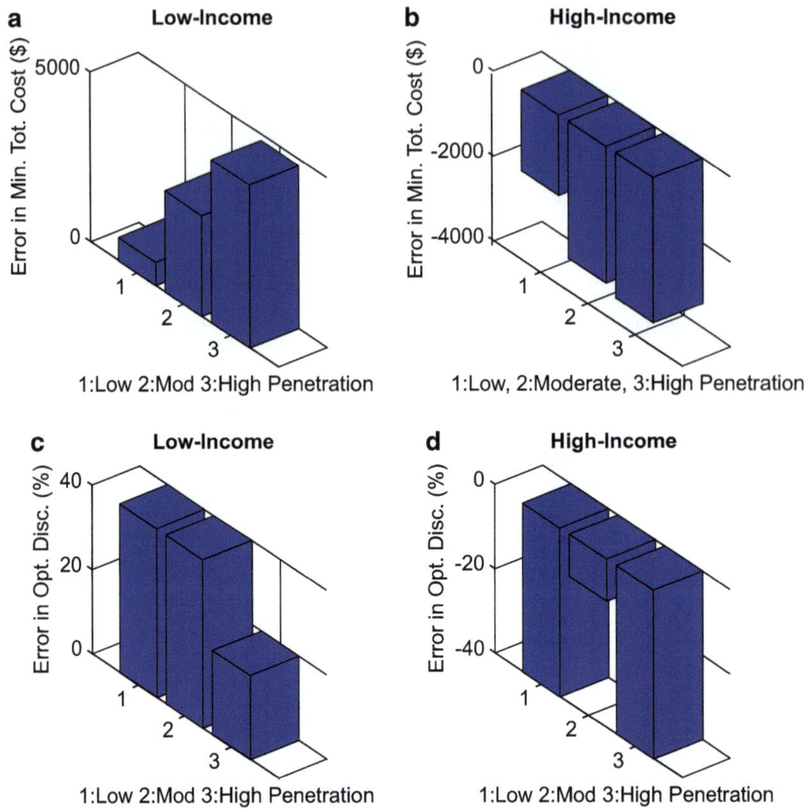

Fig. 3.13 Error in minimum total cost of problem due to considering (**a**) moderate-income instead of low-income social class, (**b**) moderate-income instead of high-income social class. Error in the optimal value of incentive due to considering (**c**) moderate-income instead of low-income social class, (**d**) moderate-income instead of high-income social class

3.6 Conclusion

In this chapter, the plug-in electric vehicle (PEV) fleet was optimally managed to enable the cooperation of responsive drivers with the generation units to provide the spinning reserve capacity of system considering the technical and social aspects of PEVs and drivers. Due to the optimal fleet management (FM) of PEVs to provide spinning reserve capacity in the normal condition and vehicle-to-grid (V2G) power in the contingency, the minimum operation cost of system decreased for any social class of drivers and for each PEV penetration level. However, the optimal FM of low-income and high-income drivers, as the most and the least responsive drivers, resulted in the least and the most minimum operation costs, respectively.

It was noticed that the optimal value of incentive needs to be updated for each social class of drivers and even for each PEV penetration level. In this regard, the high-income and low-income drivers requested the most and the least incentives, respectively.

The sensitivity analysis for the minimum operation cost of problem with respect to the value of incentive, as well as studying the effects of unrealistic modelling of drivers' social class on both minimum operation cost and incentive amount, proved that drivers' social class is a key factor in the FM of PEVs in the spinning reserve capacity provision problem.

Nomenclature

Problem parameters:	
β	Value of incentive (%) at peak period to provide the spinning reserve capacity in the normal condition and the value of discount on charging fee (%) at off-peak period after providing the V2G service in the contingency
ξ	Responsiveness probability of drivers (%)
μ	Mean of hourly Gaussian distribution function
σ	Standard deviation of hourly Gaussian distribution function
π^E	Electricity price ($/MWh)
$\alpha_1^E, \alpha_2^E, \alpha_3^E$	Emission coefficients of a generation unit in Ton/MWh^2, Ton/MW, and Ton, respectively
$\alpha_1^F, \alpha_2^F, \alpha_3^F$	Fuel cost coefficients of a generation unit in $/MWh^2, $/MWh, and $, respectively
γ^E	Emission penalty factor in $/Ton
C^{SHD}	Shut-down cost of a generation unit ($)
C^{STU}	Start-up cost of a generation unit ($)
$Cost^{INC,\ NORMAL}$	Incentive ($) paid to the responsive drivers for their availability at peak period to provide the spinning reserve capacity in the normal condition
$Cost^{INC,\ CONTINGENCY}$	Cost of discount on charging fee ($) at off-peak period offered to the responsive drivers due to their V2G service provision in the contingency at peak period
$Cost^F$	Fuel cost of generation units ($)
$Cost^E$	Greenhouse gas emissions cost of generation units ($)
$Cost^{SHD}$	Shut-down cost of a committed unit ($)
$Cost^{STU}$	Start-up cost of a decommitted unit ($)
D^{LOADs}	Load demand (excluding PEV demand) in MW
$\tilde{D}^{PEVs}$	Charging demand of responsive PEVs (MW)
$D^{PEVs,\ TOTAL}$	Total demand of PEVs (including responsive and nonresponsive PEVs) in MW
D^{PEVs}	Charging demand of nonresponsive PEVs (MW)
D^{TOTAL}	Total demand of system (MW)
DOD^{PEVs}	Depth of discharge defined for the PEVs (%)
e	Index of PEV
g, N^G	Index of generation unit and total number of generation units

(continued)

LI, MI, HI	Indices of low-income, moderate-income, and high-income drivers
MDT^G	Minimum "down" time limit of a generation unit (hour)
MUT^G	Minimum "up" time limit of a generation unit (hour)
$\tilde{N}^{PEVs}$	Expected number of responsive drivers
$N^{PEVs,\ TOTAL}$	Total number of PEVs in the area
OF	Objective function of problem
$OFFT^G$	Time interval that the generation unit has been "off" (hour)
ONT^G	Time interval that the generation unit has been "on" (hour)
P^G	Power of a generation unit (MW)
$P^{G,\ MIN}$	Minimum power limit of a generation unit (MW)
$P^{G,\ MAX}$	Maximum power limit of a generation unit (MW)
P^{PEV}	Rated power of a PEV (kWh)
$Prob^{CONTINGENCY}$	Occurrence probability of contingency (%)
RDR^G	Ramp-down rate limit of a generation unit (MW/hour).
RUR^G	Ramp-up rate limit of a generation unit (MW/hour)
s^G	Status of a generation unit, where "1" and "0" mean "on" and "off," respectively
SC	Social class of drivers
SOC^{PEVs}	State of charge of PEVs (%)
SR^{MIN}	Minimum required spinning reserve capacity of system (MW)
SR^{Gs}	Spinning reserve capacity (MW) provided by the generation units
$\tilde{SR}^{PEVs}$	Spinning reserve capacity (MW) provided by the responsive PEVs
t	Index of time (hour)
$\tilde{V2G}^{PEVs}$	V2G power (MW) provided by the responsive PEVs in the contingency
X	Binary integer as the contingency indicator (if $X = 1$)
SA parameters:	
η^{SA}	Value of coefficient to gradually decrease the temperature of molten metal
θ_0^{SA}	Initial temperature of molten metal
θ^{SA}	Current temperature of molten metal
ε^{SA}	Internal energy of molten metal
ρ^{SA}	Adaptive probability
a^{SA}	SA acceptance criterion
N^{SA}	Number of updates at each temperature
r^{SA}	A random number in the range of [0,1)

Abbreviations

DOD	Depth of discharge
ERCOT	Electric Reliability Council of Texas
FM	Fleet management
G2V	Grid-to-vehicle
LI, MI, HI	Low-income, moderate-income, high-income
PEV	Plug-in electric vehicle

(continued)

PHEV	Plug-in hybrid electric vehicle
RES	Renewable energy source
SA	Simulated annealing
SCUC	Security-constrained unit commitment
SOC	State of charge
UC	Unit commitment
V2G	Vehicle-to-grid

References

1. [Online]. Available: https://onlinelibrary.wiley.com/doi/pdf/10.1002/wene.56. Accessed on Sept 2018
2. W. Kempton et al., *A Test of Vehicle-to-Grid (V2G) for Energy Storage and Frequency Regulation in the PJM System* (Newark, Univ. Delaware, 2008)
3. "Monthly plug-in sales scorecard," Accessed on: Aug 2018. [Online]. Available: http://insideevs.com/monthly-plug-in-sales-scorecard. Accessed on Sept 2018
4. Clean Energy Ministerial, "Electric vehicles initiative," Accessed on: Aug 2018. [Online]. Available: http://cleanenergyministerial.org/Our-Work/Initiatives/Electric-Vehicles. Accessed on Sept 2018
5. [Online]. Available: https://www.smartgrid.gov/the_smart_grid/plugin_electric_vehicles.html. Accessed on Sept 2018
6. [Online]. Available: https://www.bloomberg.com/features/2016-ev-oil-crisis/. Accessed on Jan 2019
7. [Online]. Available: http://theconversation.com/affordable-batteriesfor-green-energy-are-closer-than-we-think-28772. Accessed on Sept 2018
8. M. Rahmani-Andebili, M.F. Firuzabad, M. Moeini-Aghtaie, Chapter 11: Optimal incentive plans for plug-in electric vehicles, in *Electric Distribution Network Planning*, (Springer, 2018), pp. 299–320
9. M. Rahmani-Andebili, G.K. Venayagamoorthy, SmartPark placement and operation for improving system reliability and market participation. Electr. Power Syst. Res. (Elsevier) **123** (6), 21–30 (2015)
10. M. Rahmani-Andebili, Spinning reserve supply with presence of electric vehicles aggregator considering compromise between cost and reliability. IET Gener. Transm. Distrib. **7**(12), 1442–1452 (2013)
11. M. Rahmani-Andebili, G.K. Venayagamoorthy, Investigating effects of changes in power market regulations on demand-side resources aggregators, in *IEEE Power & Energy Society General Meeting*, Denver, pp. 1–5, 20–30 July 2015
12. P. Aliasghari, B. Mohammadi-Ivatloo, M. Alipour, M. Abapour, K. Zare, Optimal scheduling of plug-in electric vehicles and renewable microgrid in energy and reserve markets considering demand response program. J. Clean. Prod. **186**, 293–303 (2018)
13. I. Pavic, T. Capuder, I. Kuzle, Value of flexible electric vehicles in providing spinning reserve services. Appl. Energy **157**, 60–74 (2015)
14. M. Alipour, B. Mohammadi-Ivatloo, M. Moradi-Dalvand, K. Zare, Stochastic scheduling of aggregators of plug-in electric vehicles for participation in energy and ancillary service markets. Energy **118**, 1168–1179 (2017)
15. M. Rahmani-Andebili, M.F. Firuzabad, An adaptive approach for PEVs charging management and reconfiguration of electrical distribution system penetrated by renewables. IEEE Trans. Ind. Inf. **14**(5) (2018)

16. H. Saadat, *Power System Analysis* (McGraw-Hill, New York, 2009)
17. D. Newbery, The economics of electric vehicles, in *EPRG and Imperial College London, E&E Seminar*, Cambridge, U.K., Jan 2013. [Online]. Available: http://www.eprg.group.cam.ac.uk/wpcontent/uploads/2013/01/EEJan13_EconomicsEVs.pdf
18. U.S. Energy Information Administration (EIA). [Online]. Available: http://www.eia.gov/todayinenergy/detail.cfm?id=9310. Accessed on Aug 2018
19. [Online]. Available: http://www.ferc.gov/market-oversight/mkt-electric/overview.asp. Accessed on Aug 2018

Chapter 4
Robust Operation of a Reconfigurable Electrical Distribution System by Optimal Charging Management of Plug-In Electric Vehicles Considering the Technical, Social, and Geographical Aspects

4.1 Introduction

The analysis of failure statistics of utilities indicates that electrical distribution system makes the greatest unavailability of supply to its customers, although the forced outages in that have a restricted effect on the end-user consumers compared with the ones in transmission and generation systems [1, 2]. On the other hand, distribution system has the most share of energy loss compared to different parts of a power system, due to its radial structure and the higher ratio of current to voltage [1]. Optimal distribution system reconfiguration (DSR) is one of the effective measures that can decrease the system power loss and enhance its reliability considering the technical constraints.

Due to the economic and environmental benefits of application of plug-in electric vehicles (PEVs), these vehicles are gaining popularity in the auto-markets [3, 4] because the PEVs can be charged by the renewable energy sources (RESs) as the clean and free energy sources. Figure 4.1 shows the battery pack cost projections ($/kWh) done by some organizations and manufactures including the US Energy Information Administration (EIA), Advanced Battery Technologies Inc., Tesla, General Motors (GM), and Nissan [5]. As can be seen, all of them predict that the battery pack cost has a descending trend in the future that will positively affect the electric vehicle market. Figure 4.2 illustrates the global electric vehicle sales (in number and percentage) including the PEV and battery electric vehicle in the period of 2010–2017 [6]. In addition, this figure shows the electric vehicle sales growth (percent) in each year.

Figure 4.3 displays the electric vehicle index development (market electric vehicle index and industry electric vehicle index) of some of the selected countries for the period of 2014–2018 [6]. In this figure, the scores are out 5. In addition, the abovementioned development indices of selected countries and their rank in 2018 are shown in Fig. 4.4 [6]. As can be noticed from these figures, China is performing better than the other countries on both market and industry electric vehicle indices.

M. Rahmani-Andebili, *Planning and Operation of Plug-In Electric Vehicles*,
https://doi.org/10.1007/978-3-030-18022-5_4

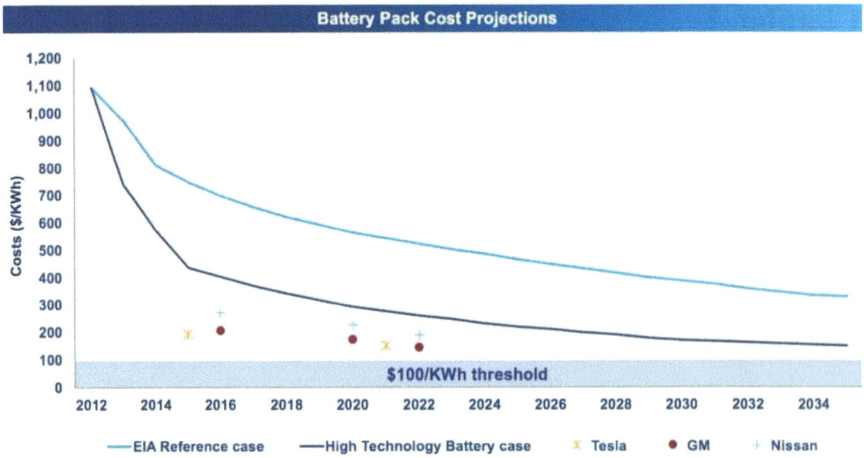

Fig. 4.1 Battery pack cost projections ($/kWh) performed by some manufactures and organizations [5]. (Source: Wood Mackenzie; EIA; Manufacturer sources)

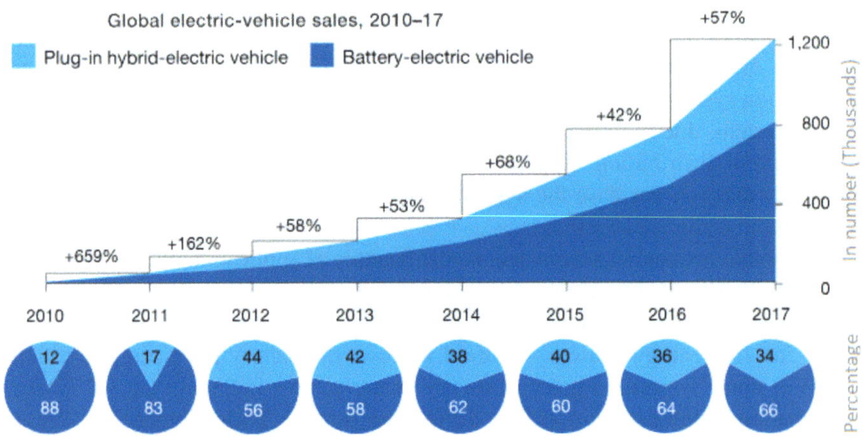

Fig. 4.2 The global electric vehicle sales including PEV and battery electric vehicle (in number and percentage) in the period of 2010–2017 [6]

In fact, the electric vehicle penetration grew from 0.3% to 2.2%, and the industry share of electric vehicle rose remarkably in China. Both France and Germany had relatively slow improvements on both market- and industry-sides in the period of 2014–2018. In other words, the electric vehicle market share in these countries increased from 0.7% to 1.7% and from 0.4% to 1.5%, respectively. Surprisingly, the Italy's market-side performance decreased, while there was a slow increase on the industry-side index. Moreover, Japan lost ground on the industry-side, and there was almost no improvement on the market-side. Norway raised the electric vehicle penetration from 11% to 32% in just 4 years. The USA had also comparatively slow

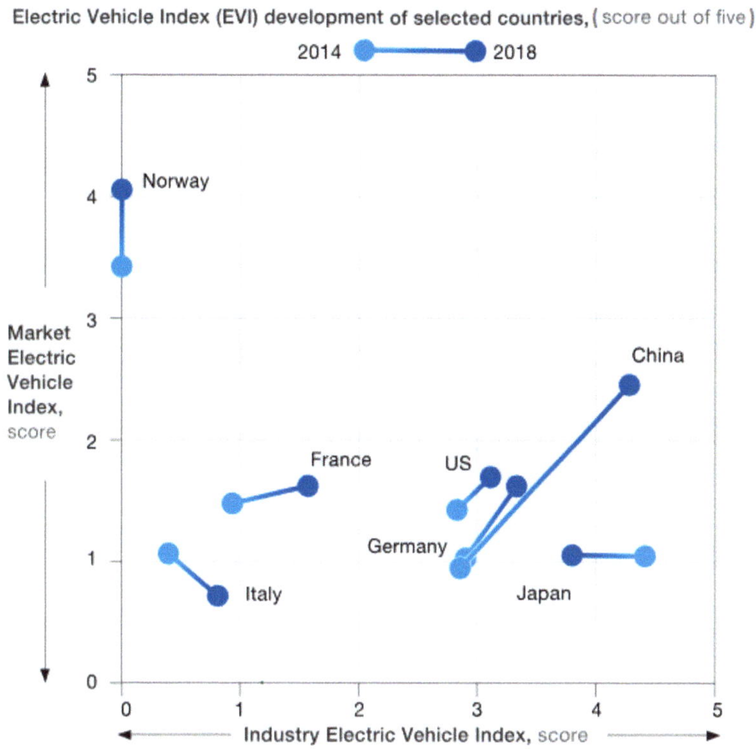

Fig. 4.3 Electric vehicle index development of some of the selected countries for the period of 2014–2018 [6]

advances on both market- and industry-sides. Figure 4.4 presents the rank of selected countries about the market and industry electric vehicle indices.

The uncontrolled charging/discharging pattern of drivers is a serious issue that might put the power system at risk. In this regard, the implementation of an efficient fleet management (FM) can result in the power loss reduction and reliability enhancement. On the other hand, PEVs, as the moveable loads or mobile energy storage systems (ESSs), can modify the load distribution of system and moderate the side effects of variable power of RESs. Due to the installation of RESs in the electrical distribution grid, ESSs play a critical role in the power system planning and operation. In fact, ESSs can be integrated with RESs to mitigate their power variability, minimize the system power loss, improve the system voltage profile, and mitigate the congestion of feeders.

Energy management problem in a smart grid has been investigated in [7, 8]. Operation of an electrical distribution system under high penetration of RESs and massive presence of PEVs is one of the major challenges of power system studies. Thus, such a system needs to be smartly and optimally operated, especially when it is under dynamic reconfiguration and includes ESSs.

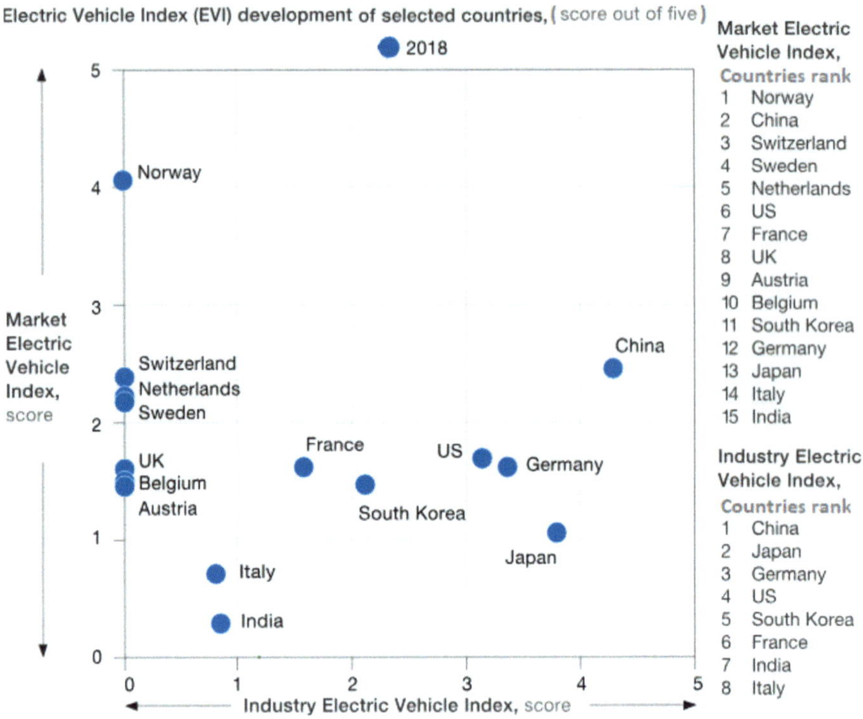

Fig. 4.4 Electric vehicle index development of selected countries and their rank in 2018 [6]

In [9–13], the operation problem of a distribution system has been studied considering the DSR problem and presence of RESs. In these studies, the presence of PEVs and ESSs has been ignored. In addition, the problems have been solved in a day-ahead approach and once for the whole period. In other words, the optimization problems have not considered the real-time variations of RES power. Moreover, the stochastic nature of RESs has been ignored in [9, 10], and the problem has been solved deterministically.

In [14], the operation problem includes DSR considering some charging patterns for the PEVs. However, the presence of RESs and ESSs has been disregarded. In [15], the optimization problem does not have any dynamic and adaptability features, as well as the behavior of drivers has not been modelled. The shortcomings of the existing studies are summarized as follows:

- In some studies, the presence of PEVs, RESs, ESSs, or all of them has been ignored in the problem.
- In the studies that include PEV fleet and RESs, the problem has been solved in a day-ahead approach and once for the whole period. Thus, the optimization problem is not updated throughout the operation period.
- In those papers that include PEVs, the behavior of drivers has been modelled using a simple approach.

In [16], the behavior of drivers has been modelled in the DSR problem; however, the presence of ESSs and the grid-to-vehicle (G2V) capability of PEVs have been ignored.

The present study intends to investigate the operation problem of a distribution system adequately supplied by the RESs. In other words, the daily energy generated by the RESs is sufficient to supply the system demand; however, the economical and reliable operation of system is desirable. Moreover, the system is equipped with the ESSs, dynamically reconfigured, and hosts the PEVs with the capabilities of vehicle-to-grid (V2G) and G2V services. The objective function of problem is to minimize the daily operation cost of system that includes system energy loss cost, system risk cost, ESS operation cost, DSR cost, and PEVs' FM cost.

In this chapter, a stochastic model predictive control (MPC) technique is applied to stochastically, adaptively, and dynamically manage the variability and uncertainties associated with the probabilistic power of RESs and drivers' behavior. In addition, the drivers' behavior is modelled based on their social class, their hourly distance from a charging station (CHS), and the value of incentive offered by the system operator to provide the G2V or V2G services at the suggested CHSs and intervals. In this regard, the geography and traffic speed limits of area are taken into account.

One of the contributions of study is comprehensively investigating the operation problem of an electrical distribution system from the economic and reliability points of view. The others include:

- Application of a stochastic, dynamic, and adaptive approach to solve the operation problem of electrical distribution system highly penetrated by the RESs
- Modelling the behavior of PEVs' owners based on their social class
- Studying the effects of drivers' behavioral models on the optimal scheme of FM at each CHS, optimal switching pattern of ESSs, optimal configuration of network, and problem objective function from economic and reliability points of view
- Investigating the effects of proposed approach on the robustness of problem outcomes with respect to different values of prediction error

In the following, Sects. 4.2, 4.3, and 4.5 are concerned with the proposed approach, problem formulation, problem simulation and numerical study, and conclusion, respectively.

4.2 Proposed Approach

4.2.1 Modelling the Geographical, Social, and Technical Aspects of Problem

It is obvious that drivers have dissimilar reactions with respect to the value of incentive when they are called by the system operator to provide the G2V or V2G services at a CHS. This hypothesis is more logical especially when the drivers have different distances from the suggested CHS. Herein, two incentives are considered for the drivers including the discount on charging fee (for the G2V service provision) and the drivers' profit (as the percentage of electricity price for the V2G service provision).

The incentive-based behavioral models of driver have been presented in [17] that the models have a dependency just on the value of incentive and driver's social class. However, in this study, the models are extended to consider the driver's distance (meter) from the CHS, in addition to the abovementioned parameters. The drivers are grouped in three different social classes based on their income level, namely, low-income, moderate-income, and high-income. The low-income and high-income drivers are considered as the most and the least responsive drivers, since they are more and less likely to be interested in the proposed incentive to cooperate with the system operator. Therefore, a power function and an exponential function are used to model their behavior with respect to value of incentive, respectively. Moreover, a linear function is applied to model the behavior of moderate-income drivers with respect to value of incentive. Regarding the dependency on distance, a linear model is applied for each model with the parameter of α, as the distance of driver from the CHS.

The responsiveness probability functions of a low-income, moderate-income, and high-income driver with respect to the value of incentive and his/her distance from the CHS for the ranges of $0 \le \beta \le 100$ percent and $0 \le \alpha \le 1000$ meter are presented in Fig. 4.5 and Table 4.1 in Eqs. (4.1, 4.2, and 4.3). In Table 4.1, *e*, *LI*, *MI*, *HI*, and *t* are the indices of PEV, low-income driver, moderate-income driver, high-income driver, and time, respectively. As can be seen in Fig. 4.5, the whole 3-D space is rationally covered using these three models to consider all the possibilities of driver's reaction. Herein, some logics can be realized from the graphs and Table 4.1, presented as follows:

- The driver, with any distance from the suggested CHS, will not refer to the CHS to charge/discharge his/her vehicle if there is no incentive ($\beta = 0\%$).
- The driver, with the nearest distance from the suggested CHS ($\alpha \le 100$ meters), will charge/discharge his/her vehicle through the CHS if it is free ($\beta = 100\%$).
- The driver, with any incentive, will not charge/discharge his/her vehicle through the suggested CHS if the CHS is farther than the maximum specified distance ($\alpha > 1000$ meters).

Table 4.1 Responsiveness probability of a low-income (LI), moderate-income (MI), and high-income (HI) driver with respect to the value of incentive (β) and his/her distance from the CHS (α)

Social class	Driver's responsiveness probability (%)	
Low-income drivers	$\xi_{e,LI,\alpha,\beta,t} = 100 \times \left(1 - \frac{\alpha_e}{1000}\right) \times \left(\frac{\beta}{100}\right)^{0.3}$	(4.1)
Moderate-income drivers	$\xi_{e,MI,\alpha,\beta,t} = 100 \times \left(1 - \frac{\alpha_e}{1000}\right) \times \left(\frac{\beta}{100}\right)$	(4.2)
High-income drivers	$\xi_{e,HI,\alpha,\beta,t} = 100 \times \left(1 - \frac{\alpha_e}{1000}\right) \times exp\left(10 \times \left(\frac{\beta}{100} - 1\right)\right)$	(4.3)

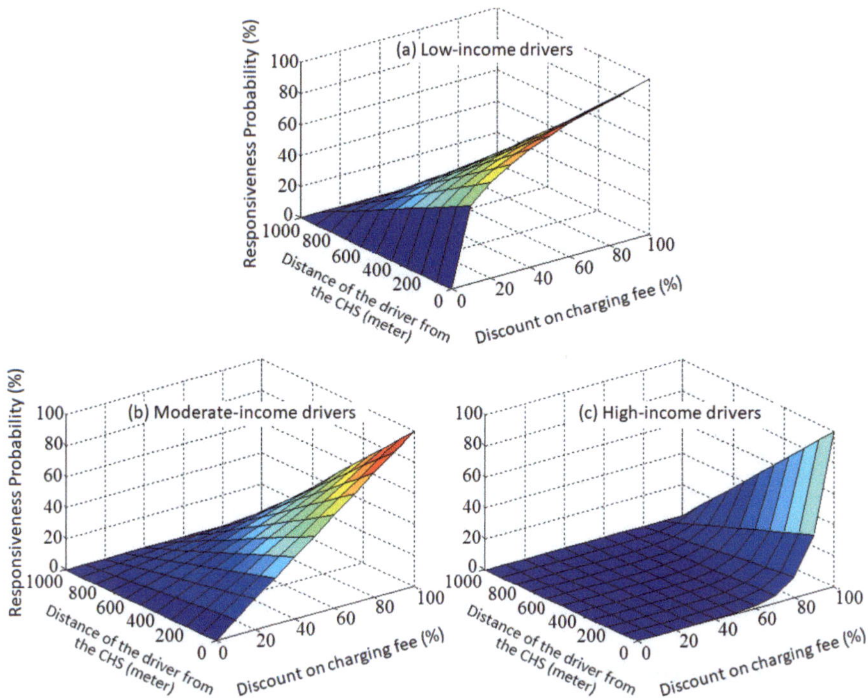

Fig. 4.5 Responsiveness probability of a driver with different social classes with respect to the value of incentive and his/her distance from the CHS for the ranges of $0 \leq \beta \leq 100$ percent and $0 \leq \alpha \leq 1000$ meter

The expected number of responsive drivers ($\tilde{N}^{PEVs}$) is determined using Eq. (4.4), where SC and $N^{PEVs,\ TOTAL}$ are the index of social class and the total number of PEVs in the area, respectively.

$$\tilde{N}^{PEVs}_{SC,\alpha,\beta,t} = \sum_{e=1}^{N^{PEVs,\,TOTAL}} \frac{\xi_{e,SC,\alpha,\beta,t}}{100}, \quad \forall SC \in \{LI, MI, HI\}, \ \ \forall \alpha \in \{0, 100, \ldots, 1000\}, \\ \forall \beta \in \{0, 10, \ldots, 100\}, \ \ \forall t \in \{1, \ldots, T\} \tag{4.4}$$

Figure 4.6 illustrates the hourly space-time driving routes of PEVs in the typical day. These driving patterns have been randomly generated by a computer while considering the minimum and maximum traffic speed limits and the real latitude and longitude of area in Tehran using Google Maps. By looking at Fig. 4.6, it can be realized that the vehicles are immobile between 1st–7th and 23rd–24th hours of the day, since they are parked during these time intervals. In this study, each vehicle is the representative of 50 PEVs, thus 500 PEVs exist in the area.

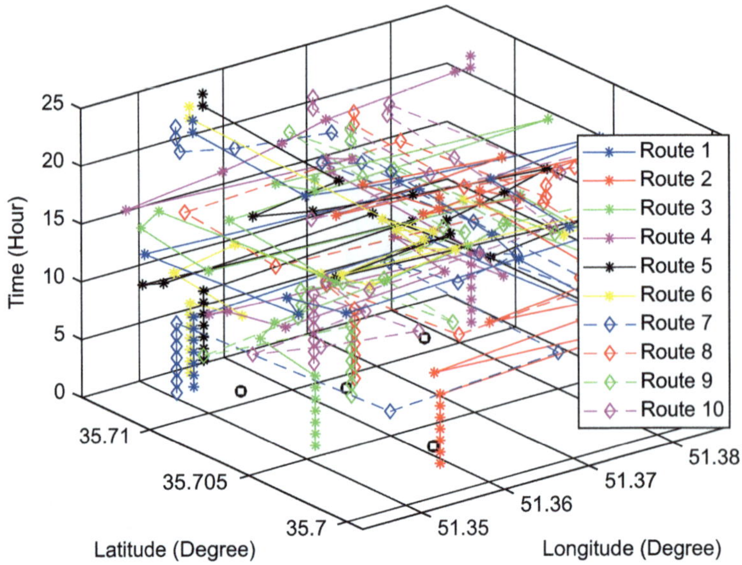

Fig. 4.6 Hourly space-time driving routes throughout the typical day

4.2.2 Stochastic Model Predictive Control

A combination of stochastic approach and MPC technique is applied in this study to deal with the uncertainty and variability issues of problem related to the probabilistic power of RESs and drivers' behavior. By application of stochastic MPC, the problem is guaranteed to be solved stochastically, adaptively, and dynamically.

4.2.2.1 Model Predictive Control

MPC can control a multivariable constrained system by taking the control actions from the solution of an online optimization problem and predicting the system's behavior repetitively [18]. Figure 4.7 shows the structure of MPC for the 1 hour time step and optimization time horizon of $t + 1, \cdots, t + n^{\tau}$, where n^{τ} is six in this study [19, 20].

Considering a long optimization time horizon creates a vast vision for the optimization and forecasting time horizons; however, it will need more computations and result in the computational latency. On the other hand, short optimization time horizon needs less computations; nonetheless, the outcomes might be far from the optimal results, due to the local optimization. Thus, considering an appropriate length for the optimization time horizon is advantageous.

In the application of MPC, the value of cost-to-go function (J_t), known as Hamilton-Jacobi-Bellman's equation, is minimized for the optimization time horizon ($t + 1, \cdots, t + 6$); however, just the variables of next time step ($t + 1$) are accepted

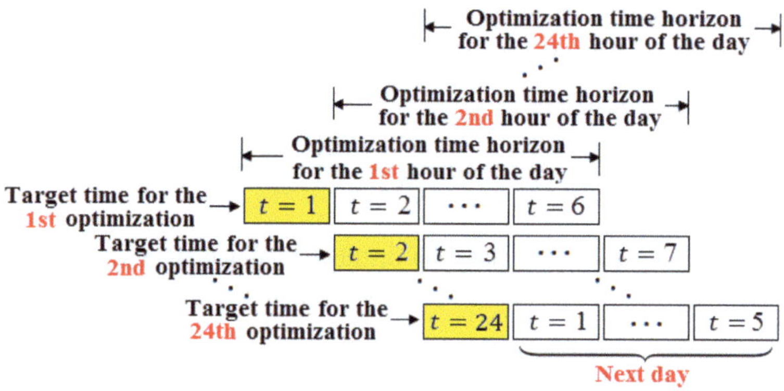

Fig. 4.7 Schematic of MPC with six time step optimization horizons [19, 20]

as the decision variables, as can be seen in Fig. 4.7 and Eq. (4.5). In fact, the value of J_t is the sum of utility functions (U) over the optimization time horizon ($t + 1, \cdots, t + n^{\tau}$) weighted by a damping multiplier ($z^{\tau - 1}$).

$$min\{J_t\} = min\left\{\sum_{\tau=1}^{n^{\tau}} z^{\tau-1} \times U_{t+\tau}\right\}, \quad \forall t \in \{1, \ldots, T\} \tag{4.5}$$

The value of multiplier descends over the optimization horizon, since the absolute value of discount factor (z) is less than unit ($0 \leq z < 1$). Choosing a large value for the discount factor ($z \approx 1$) means considering the effect of states of far future on the decision variables; however, selecting a small value for the discount factor ($z \ll 1$) results in focusing on the states of near future. Therefore, a suitable value is needed to achieve the satisfactory outcomes.

Next, as can be seen in Fig. 4.7, the optimization time horizon is updated and changed to $\{t + 2, \cdots, t + 7\}$. Then, the value of updated cost-to-go function (J_{t+1}) is optimized, and the variables of next time step ($t + 2$) are accepted as the decision variables. This procedure that demonstrates the dynamic and adaptability characteristics of MPC is repeated for every time step (hour) of operation period ($1, \cdots, 24$).

4.2.2.2 Stochastic Approach

The stochastic approach includes forecasting the value of uncertain states (wind speed and state of charge (SOC) of PEV fleet) over the optimization time horizon and addressing the predictions uncertainty.

Probabilistic SOC of PEV fleet The uncertainty of SOC of PEV fleet is modelled applying a Gaussian distribution function, presented in Eq. (4.6). In other words, the hourly value of SOC of PEV fleet is modelled using two parameters of Gaussian distribution, that is, mean (μ) and standard deviation (σ).

$$f_t^{SOC} = \frac{1}{\sigma_t \sqrt{2\pi}} e^{-\frac{\left(SOC_t^{PEVs} - \mu_t\right)^2}{2\sigma_t^2}} \tag{4.6}$$

Figure 4.8a illustrates the Gaussian distribution functions associated with the hourly SOC of PEV fleet [21]. Herein, some logics can be noticed presented as follows:

- The SOC of PEV fleet has a descending trend throughout the day, since the PEVs are losing their charge.
- The SOC of PEV fleet for the intervals of 1–7 and 22–24 has small standard deviation, since the majority of PEVs are parked at CHSs.
- The SOC of PEV fleet for the periods of 8–21 has large standard deviation, since the PEVs are on the streets.

The Gaussian distribution function related to the SOC of PEV fleet at 21st hour is shown in Fig. 4.8b. Based on the Gaussian distribution function given for each time step (1 hour), four values for the SOC of PEV fleet (corresponding to $\mu - 2\sigma$, $\mu - \sigma$, $\mu + \sigma$, and $\mu + 2\sigma$) are estimated. These estimated values have the occurrence probabilities of 15.9%, 34.1%, 34.1%, and 15.9%, respectively.

Probabilistic wind power The other uncertain states of problem include the power of wind turbines (or wind speed) over the optimization time horizon ($t + 1, \cdots, t + n^{\tau}$). These uncertain states are modelled using Weibull random variable. The probability density function of a Weibull random variable is given in Eq. (4.7), where k and η are the shape and scale parameters of distribution function, respectively [22]. Figure 4.9a, c, d, and e shows the hourly Weibull distribution functions associated with four wind turbines. In addition, Fig. 4.9b illustrates the Weibull distribution function related to the first wind turbine at 22nd hour of the day.

$$f_x^{Weibull} = \begin{cases} \frac{k}{\eta}\left(\frac{x}{\eta}\right)^{k-1} e^{-\left(\frac{x}{\eta}\right)^k} & x \geq 0 \\ 0 & x < 0 \end{cases}, \quad k,\eta > 0 \tag{4.7}$$

To model the uncertainty of predictions, three values for the hourly power of each wind turbine, corresponding to $0.5 \times \eta$, η, and $2 \times \eta$, are estimated, as can be seen in Fig. 4.9b. These three estimated values have distinct occurrence probabilities in accordance with the shape and scale parameters of related Weibull distribution.

Scenario generation and selection All the possible scenarios for the problem are defined using all the estimated uncertain states over the optimization time horizon. It is obvious that considering more scenarios will result in more accurate outcomes for the problem; however, it may lead to an unmanageable optimization problem, since

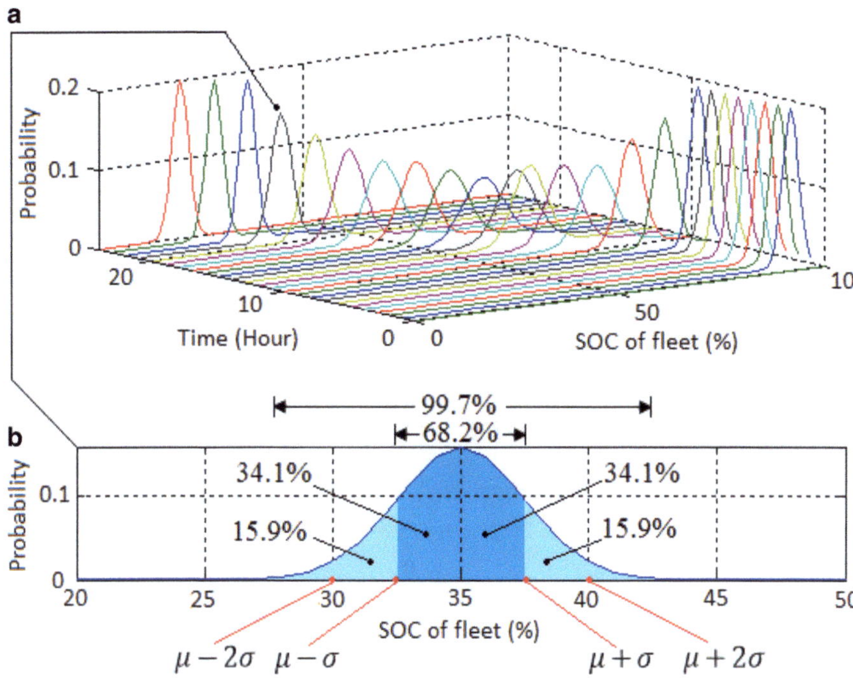

Fig. 4.8 (**a**) Gaussian distribution functions related to hourly SOC of PEV fleet. (**b**) Gaussian distribution function of SOC of PEV fleet at 21st hour of the day [21]

the optimization problem cannot be solved in the desirable time (1 hour), due to the large number of scenarios. Therefore, to avoid dealing with an unmanageable optimization problem, 20 scenarios are randomly selected using roulette wheel mechanism (RWM). By using RWM, it is guaranteed that the most probable and diverse scenarios are chosen.

In the stochastic MPC, the cost-to-go function (J) presented in Eq. (4.5) is changed to stochastic cost-to-go function (J^{Stoch}), as can be seen in Eq. (4.8). Herein, S is the set of all the possible scenarios that include scenarios associated with the SOC of PEV fleet (S^{SOC}) and power (or wind speed) of each wind turbine (S^{W}). The number of scenarios covered by S at each time step is $N^{(4)\times(n^{\tau})}$, where N is the number of uncertain states, "4" is the number of estimated values for each uncertain state, and n^{τ} is the optimization time horizon. In addition, S^{RWM} is the scenarios chosen by RWM, and p is the occurrence probability of a scenario.

$$J_t^{Stoch} = \sum_{s \in S^{RWM}} \frac{p_{t,s}}{100} \times J_{t,s}, \quad S^{RWM} \epsilon S, \ S = \{S^{SOC}, S_1^W, \ldots, S_w^W\} \tag{4.8}$$

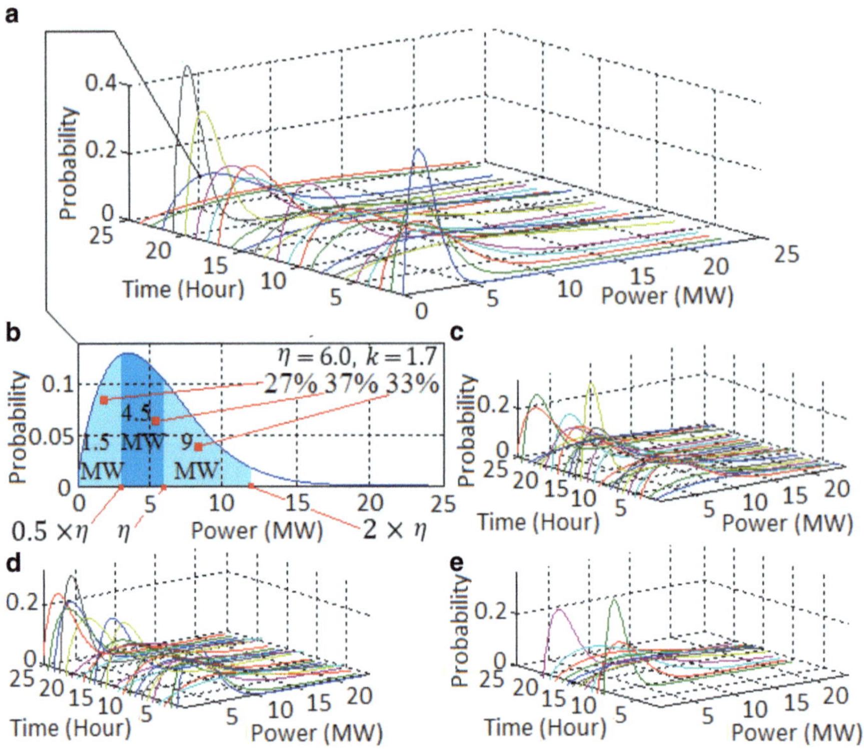

Fig. 4.9 (**a**), (**c**), (**d**), and (**e**): hourly Weibull distribution functions associated with four wind turbines. (**b**): The Weibull distribution function related to the first wind turbine at 22nd hour

4.2.3 Optimization Technique

In this study, an improved genetic algorithm (IGA) is applied to solve the optimization problem. Other algorithms could be used in this problem; however, the powerful search capability of IGA in the complex and nonlinear environments is its competences compared to other algorithms.

The IGA is applied at every time step of operation period (1, $\cdots$, 24) for the updated optimization time horizon (t + 1, $\cdots$, t + 6). Thus, the defined chromosome includes the problem variables of all the time steps of optimization time horizon. The problem variables at each time step include the status of feeders' switches for being open or close ($\varphi^{SW} \in \{0, 1\}$), the mode of ESSs for being in charging/discharging/idles status ($\varphi^{ESS} \in \{-1, 0, 1\}$), the status of CHSs to provide G2V or V2G services ($\varphi^{G2V - V2G} \in \{0, 1\}$), and the value of incentive at each CHS ($\beta^{CHS} \in \{0, 10\%, \ldots, 100\%\}$). These variables need one gene, two genes, one gene, and four genes for each time step, respectively. Coding the value of incentive from decimal number (β^{CHS}) to the binary integer needs four binary digits (γ_0^{CHS}, γ_1^{CHS}, γ_2^{CHS}, and γ_3^{CHS}) because it is changed from 0% to 100% with a 10% step.

The inverse value of operation cost over the optimization time horizon is defined as the fitness of a chromosome, and then the IGA tries to improve the fitness of chromosomes of population.

In this study, the mutation probability of genes (θ^M) is adaptively controlled to expedite the optimization process, as can be seen in Eq. (4.9). The $\Delta Fitness$ is the value of difference between the average fitness of chromosomes of current population and the previous ones. Based on this, the initial value of mutation probability is close to 50% (because $\Delta Fitness$ is large); however, when the optimization procedure is in its final stages ($\Delta Fitness$ is small), it is descended to almost zero. Adding this characteristic to the GA results in the global searching in the initial stages and local searching in the final stages.

$$\theta^M = 50 \times e^{-\frac{1}{\Delta Fitness}} \tag{4.9}$$

4.3 Problem Formulation

4.3.1 Objective Function

The objective function of problem (OF) is to minimize the daily operation cost of distribution grid that includes sum of the hourly stochastic cost-to-go functions. The stochastic cost-to-go function (J^{Stoch}), cost-to-go function (J), and utility function (U) have been presented in Eqs. (4.8), (4.5), and (4.11), respectively. As can be seen in Eq. (4.11), the utility function, as the sum of cost terms, includes energy loss cost ($Cost^{LOSS}$), energy not supplied (ENS) cost ($Cost^{ENS}$), DSR cost ($Cost^{DSR}$), FM cost ($Cost^{FM}$), and ESS operation cost ($Cost^{ESS}$).

$$OF = min\left\{\sum_{t \in T} J_t^{Stoch}\right\} \tag{4.10}$$

$$U_t = Cost_t^{LOSS} + Cost_t^{ENS} + Cost_t^{DSR} + Cost_t^{FM} + Cost_t^{ESS} \tag{4.11}$$

4.3.2 Cost Terms

4.3.2.1 Energy Loss Cost

The hourly energy loss cost of branches of feeders is presented in Eq. (4.12). In addition, the hourly value of active power loss of a branch of a feeder in MW (P^{LOSS}) is presented in Eq. (4.13). Herein, π^E, $R^{p.\ u.}$, $|I^{p.\ u.}|$, and MVA^{BASE} are the electricity price ($/MWh), resistance of branch (p.u.), magnitude of current flowing through the branch (p.u.), and base MVA defined for the entire system, respectively.

$$Cost_t^{LOSS} = \sum_{f \in F} \sum_{br \in BR} P_{f,br,t}^{LOSS} \times \pi^E, \ F = \{1, \ldots, N^F\}, \ BR = \{1, \ldots, N^{BR}\} \tag{4.12}$$

$$P_{f,br,t}^{LOSS} = R_{f,br}^{p.u.} \times \left| I_{f,br,t}^{p.u.} \right|^2 \times MVA^{BASE} \tag{4.13}$$

4.3.2.2 ENS Cost

The hourly ENS cost of system is presented in Eq. (4.14). Herein, π^{ENS} is the hourly damage ($/MWh) imposed to the customers because of a contingency like occurring a fault in the branch. Additionally, π^{ENS} has different values at different periods including valley period ($t \in 1, \ldots, 8$), off-peak period ($t \in 9, \ldots, 16$), and peak period ($t \in 17, \ldots, 24$).

The hourly value of expected ENS (EENS) of system (MWh) is calculated using Eq. (4.15) [2, 23]. This reliability index, as the risk level of system, depends on the failure rate of each branch of each feeder (λ) in failure/year, the failure locating and DSR duration (T^{FL-DSR}) in hour, and the failure repairing duration (T^{FR}) in hour. Moreover, LNS^{FL-DSR} (MW) is the value of load not supplied during the locating and DSR intervals, and LNS^{FR} (MW) is the value of load not supplied during the repairing period.

$$Cost_t^{ENS} = EENS_t \times \pi_t^{ENS} \tag{4.14}$$

$$EENS_t = \frac{1}{365} \sum_{f \in F} \sum_{br \in BR} \lambda_{br} \times \begin{pmatrix} T^{FL-DSR} \times \sum_{b \in B} LNS_{f,b,t}^{FL-DSR} \\ +T^{FR} \times \sum_{b \in B} LNS_{f,b,t}^{FR} \end{pmatrix} \tag{4.15}$$

4.3.2.3 DSR Cost

The hourly DSR cost includes the switching cost of initially open and close switches, as can be seen in Eq. (4.16). This cost term is related to the degradation cost of switches due to their switching. Herein, π^{SW} ($) indicates the feeder's switching cost for once, φ^{SW} is the status of switch (open or close), and $\acute{\varphi}^{SW}$ is mathematically described in Eq. (4.17).

$$Cost_t^{DSR} = \sum_{sw \in SW} \acute{\varphi}_{sw,t}^{SW} \times \pi^{SW}, \quad SW = \{1, \ldots, N^{SW}\} \tag{4.16}$$

$$\acute{\varphi}_{sw,t}^{SW} = \begin{cases} 0 & \varphi_{sw,t-1}^{SW} = \varphi_{sw,t}^{SW} \\ 1 & \varphi_{sw,t-1}^{SW} \neq \varphi_{sw,t}^{SW} \end{cases} \tag{4.17}$$

4.3.2.4 FM Cost

The hourly FM cost includes the hourly incentive (β^{CHS}) paid to the drivers for their G2V/V2G service provision at different CHSs ($chs \in CHS$) considering the demand/power of G2V/V2G of their vehicles (P^{G2V} or P^{G2V}) and the social class of drivers ($SC \in \{LI, MI, HI\}$), as can be seen in Eq. (4.18). The $\varphi^{G2V-V2G}$ is a binary integer that indicates the status of vehicles to provide G2V or V2G service. The binary value of incentive (γ_0^{CHS}, γ_1^{CHS}, γ_2^{CHS}, and γ_3^{CHS}) is changed to decimal number using Eq. (4.19).

As can be seen in Eq. (4.20), the value of G2V of PEVs at a CHS depends on the SOC level of PEVs, the power of battery of PEVs (P^{PEV}), and the number of responsive drivers ($\tilde{N}^{PEVs}$). The $\tilde{N}^{PEVs}$ has been defined in Eq. (4.4). In addition, the power of V2G service of PEVs at a CHS is presented in Eq. (4.21).

$$Cost_t^{FM} = \sum_{chs \in CHS} \frac{\beta_{chs,t}^{CHS}}{100} \times \left\{ \begin{array}{c} \varphi_{chs,t}^{G2V-V2G} \times P_{chs,SC,\beta,t}^{G2V} \\ +\left(1 - \varphi_{chs,t}^{G2V-V2G}\right) \times P_{chs,SC,\beta,t}^{V2G} \end{array} \right\} \times \pi^E, \tag{4.18}$$
$$CHS = \{1, \ldots, N^{CHS}\}, \quad SC \in \{LI, MI, HI\}$$

$$\beta_{chs,t}^{CHS} = \left(\gamma_{0,chs,t}^{CHS} + 2^1 \times \gamma_{1,chs,t}^{CHS} + 2^2 \times \gamma_{2,chs,t}^{CHS} + 2^3 \times \gamma_{3,chs,t}^{CHS}\right) \times 10 \tag{4.19}$$

$$P_{chs,SC,\beta,t}^{G2V} = \left(1 - \frac{SOC_t^{PEVs}}{100}\right) \times \frac{P^{PEV}}{1000} \times \tilde{N}_{chs,\alpha,SC,\beta,t}^{PEVs} \tag{4.20}$$

$$P_{chs,SC,\beta,t}^{V2G} = \left(\frac{SOC_t^{PEVs} - DOD^{PEVs}}{100}\right) \times \frac{P^{PEV}}{1000} \times \tilde{N}_{chs,\alpha,SC,\beta,t}^{PEVs} \tag{4.21}$$

4.3.2.5 ESS Operation Cost

The hourly operation cost of an ESS includes its switching cost due to changing its mode between charging, discharging, and idle statuses. This cost term is associated with the aging cost of ESSs due to their switching. The π^{ESS} indicates the switching cost of an ESS for once, φ^{ESS} is the status of ESS (charging, discharging, or idle), and $\acute{\varphi}^{ESS}$ is mathematically described in Eq. (4.23).

$$Cost_t^{ESS} = \sum_{ess \in ESS} \acute{\varphi}_{ess,t}^{ESS} \times \pi^{ESS}, \ ESS = \{1, \ldots, N^{ESS}\} \tag{4.22}$$

$$\acute{\varphi}_{ess,t}^{ESS} = \begin{cases} 0 & \varphi_{ess,t-1}^{ESS} = \varphi_{ess,t}^{ESS} \\ 1 & \varphi_{ess,t-1}^{ESS} \neq \varphi_{ess,t}^{ESS} \end{cases} \tag{4.23}$$

4.3.3 *Constraints*

4.3.3.1 Voltage Stability of Buses

The magnitude of voltage of each bus of each feeder (|V|) must be within the allowable range at each time step.

$$\underline{|V|} \leq |V_{f,b,t}| \leq \overline{|V|}, \quad \forall b \in B, \ B \in \{1, \ldots, N^B\} \tag{4.24}$$

4.3.3.2 Thermal Capacity of Branches of Feeders

The thermal capacity constraint of each branch of each feeder at each time step, as its power loading limit, is presented in Eq. (4.25).

$$|MVA_{f,br,t}| < \overline{|MVA_{f,br,t}|} \tag{4.25}$$

4.3.3.3 Radiality and Connectivity of System

These constraints are kept by applying the method presented in [16, 24]. Based on the method [16], the system configuration is defined as a graph, and then the connection matrix of graph ($\hat{A}$) is formed. In $\hat{A}$, the columns represent the system buses, and the rows indicate the system branches. Each element of $\hat{A}$ ($\hat{a}_{ij}$):

- Takes 0 if the *i*'th branch is not connected to the *j*'th bus
- Takes 1 if the *i*'th branch is connected but directed away from the *j*'th bus
- Takes −1 if the *i*'th branch is connected but directed toward the *j*'th bus

Next, a reference bus is assigned, and its corresponding column is deleted from $\hat{A}$ to form the square matrix (*A*). After that, the determinant of *A* is calculated, and if it is equal to 1 or −1, the system is known as radial and connected; otherwise, the system is not radial or connected.

4.3.3.4 Power Limit of a PEV for G2V and V2G

The G2V power (P^{G2V}) and V2G power (P^{V2G}) of each PEV at each CHS must be less than its maximum allowable value ($\overline{P^{PEV}}$) at each time step.

$$P_{e,t}^{G2V} \,\&\, P_{e,t}^{V2G} \leq \overline{P_{e}^{PEV}} \tag{4.26}$$

4.3.3.5 SOC Limits of a PEV

To prolong the lifetime of battery of a PEV, the battery should not be discharged more than the allowable depth of discharge (DOD^{PEV}). Moreover, the battery has an upper capacity limit; thus it cannot be charged beyond this limit.

$$DOD^{PEV} \leq SOC_{e,t}^{PEV} \leq 100 \tag{4.27}$$

4.3.3.6 Power Limit of an ESS

The magnitude of power of each ESS ($|P^{ESS}|$) must be less than its maximum allowable value ($\overline{P^{ESS}}$) at each time step, while it is in charging or discharging mode.

$$\left|P_{ess,t}^{ESS}\right| \leq \overline{P_{ess}^{ESS}} \tag{4.28}$$

4.3.3.7 SOC Limits of an ESS

Like battery of a PEV, each ESS has SOC constraints that must be held at each time step of optimization horizon.

$$DOD^{ESS} \leq SOC_{ess,t}^{ESS} \leq 100 \tag{4.29}$$

4.3.3.8 Zero-Energy Constraint of an ESS

This constraint guarantees that each ESS is used only as an ESS and not as a generation or load source. In other words, for each ESS, the cumulative value of charged energy must be equal to the cumulative value of discharged energy in every optimization time horizon.

$$\sum_{\tau=1}^{n^{\tau}} \varphi_{ess,t+\tau}^{ESS} \times P_{ess,t+\tau}^{ESS} = 0 \tag{4.30}$$

4.4 Problem Simulation

4.4.1 Characteristics of System and Problem

Figure 4.10 illustrates the geographic and single-line diagram of electrical distribution system that includes 1 sub-transmission substation, 4 distribution feeders (DFs 1–4), 4 initially close switches, 3 initially open switches, 47 distribution buses, 4 CHSs (installed on DF 1 (bus 2), DF 2 (bus 10), DF 3 (bus 12), DF 4 (bus 13)), 4 wind turbines (WT1–WT4), and 4 ESSs (installed on DF 3 (bus 12), DF 4 (buses 13, 15), and DF 2 (bus 11)). The hourly Weibull distribution functions associated with WT1–WT4 are shown in Fig. 4.9a, c, d, and e, respectively. In addition, the hourly Gaussian distribution functions of PEVs' SOC are illustrated in Fig. 4.8.

The value of active power demand (MW) and reactive power demand (MVAr) of each distribution bus of each feeder at 21st hour of the day (peak demand) are shown in Fig. 4.11a and b, respectively. Moreover, the hourly demand pattern (p.u.) of each distribution bus of system has been presented in [25]. The value of resistance and reactance (p.u.) of each branch of each feeder are shown in Fig. 4.12a and b,

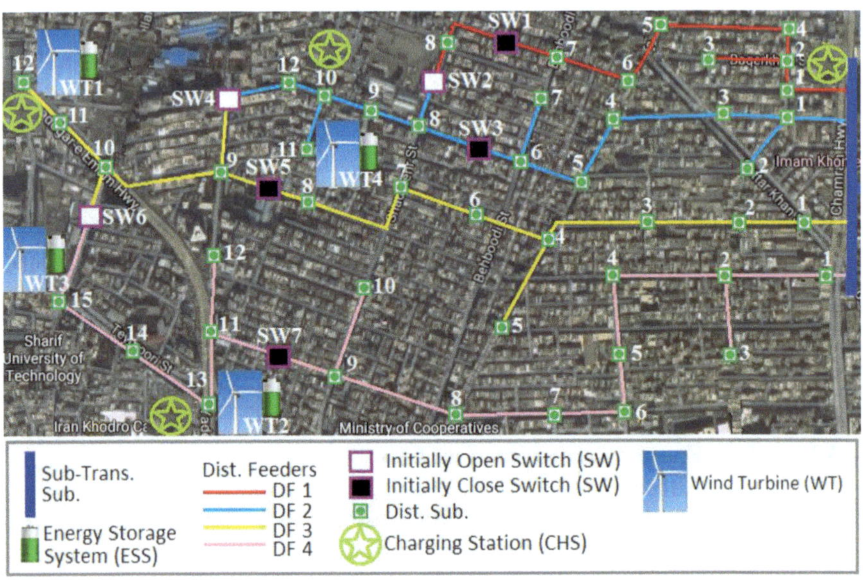

Fig. 4.10 Geographic and single-line diagram of the electrical distribution system under study

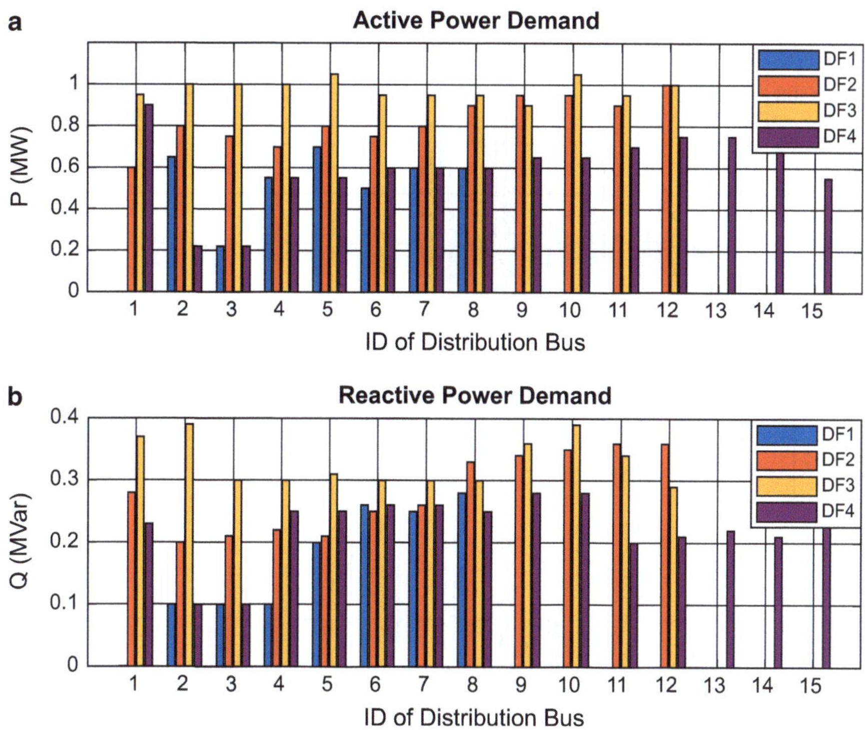

Fig. 4.11 (**a**) Value of active power demand (MW) and (**b**) reactive power demand (MVAr) of each bus of each feeder (DFs 1–4) at 21st hour of the day

respectively. Moreover, the thermal capacity (MW) of branches and switches of feeders are shown in Fig. 4.13.

Table 4.2 presents the value of parameters of problem. The model of PEVs is considered Nissan Leaf that has the rated capacity (C^{PEV}) and power ($\overline{P^{PEV}}$) of 24 kWh and 24 kW, respectively [26]. In addition, the capacity of each ESS is 5 MWh. Moreover, the hourly customer damage (π^{ENS}) due to the occurrence of a fault in a branch at valley, off-peak, and peak period are considered about 2, 10, and 5 times of electricity price, respectively.

4.4.2 Simulating the Problem in Different Scenarios

The problem is investigated in three distinct scenarios as follows:

- *First scenario:* The problem is simulated without the application of ESSs, FM, and DSR.

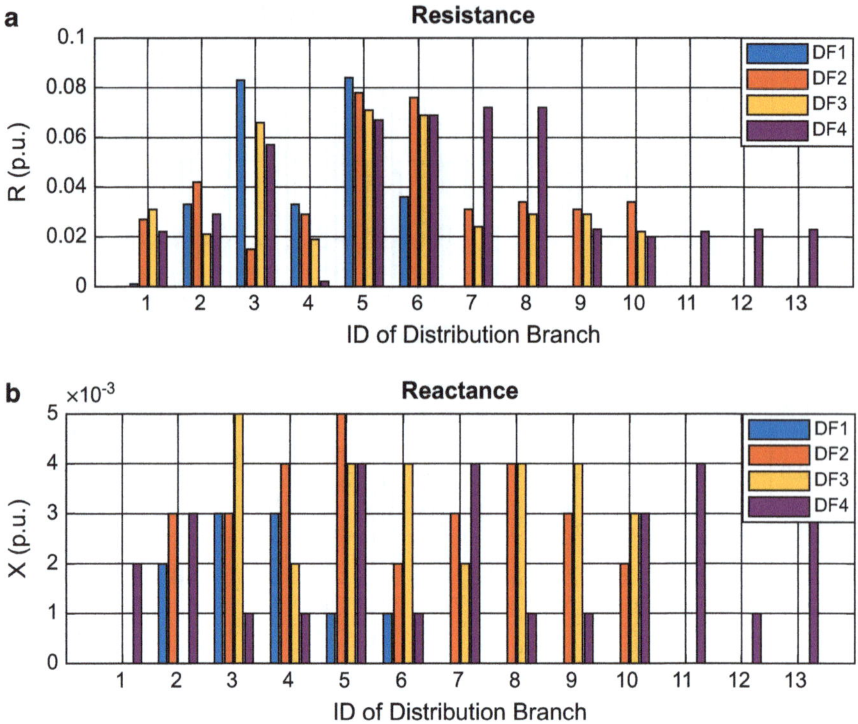

Fig. 4.12 (**a**) Value of resistance (p.u.) and (**b**) reactance (p.u.) of each branch of each feeder (DFs 1–4)

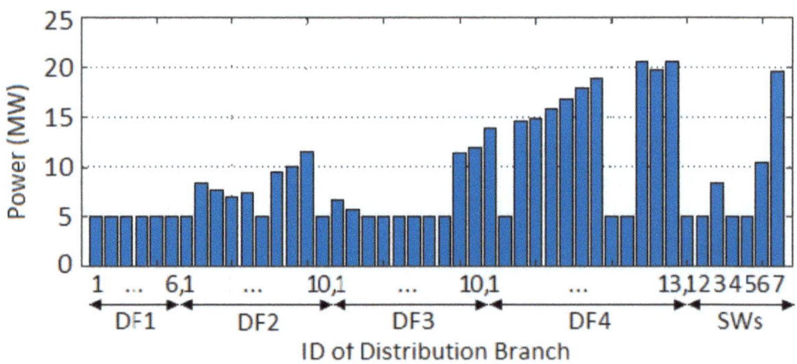

Fig. 4.13 Thermal capacity (MW) of branches and feeders' switches

- *Second scenario:* The proposed approach (presented in Sect. 4.2) is applied, and the distribution system is optimally operated.
- *Third scenario:* The robustness of proposed approach with respect to different values of prediction error is investigated.

Table 4.2 Value of problem parameters

Parameter	Value	Parameter	Value
π^{E}	$50.71/MWh	T^{FL-DSR}	0.6 hour
π^{SW}	$10	T^{FR}	1 hour
π^{ESS}	$10	C^{PEV}	24 kWh [26]
MVA^{BASE}	4 MVA	$\overline{P^{PEV}}$	24 kW [26]
$\overline{\lvert V\rvert}$	1.10 p.u.	DOD	20%
$\underline{\lvert V\rvert}$	0.90 p.u.	z	0.9
λ	0.1 failure/year	n^{τ}	6

4.4.2.1 Scenario 1

Table 4.3 presents the problem simulation results for each social class of drivers in Sc. 1 and Sc. 2. In addition, Fig. 4.14 illustrates the voltage profile of buses (p.u.) at 21st hour of the day (peak demand) before (Sc. 1) and after optimal operation of system (Sc. 2). Moreover, the value of powers (MW) flowing through the branches and feeders' switches at peak hour before (Sc. 1) and after optimal operation of system (Sc. 2) is shown in Fig. 4.15.

As can be seen in Figs. 4.14 and 4.15, there are voltage violation in buses 7–12 of DF 3 and buses 9–15 of DF 4, as well as power loading violation in some of the branches before optimal operation of system (Sc. 1).

4.4.2.2 Scenario 2

In Sc. 2, the daily operation cost, energy loss, and ENS of system decrease for any behavioral model of drivers, as can be seen in Table 4.3. However, the low-income and high-income drivers have the most and the least contributions for the system operator, since the reliability and operation cost of system have the best and the worst values, respectively. As can be seen, the optimal FM of low-income drivers results in about 37.0%, 48.0%, and 16.2% reductions for the operation cost, energy loss, and ENS, respectively.

By looking at Fig. 4.14, it is realized that the voltage profile of all buses is regulated after optimal operation of system (Sc. 2). In this condition, part of DF 4 is transferred onto DF 3, due to the optimal DSR. The direct effect of optimal DSR is observed in the sudden voltage improvement of Buses 9–12 of DF 3, since this part of DF 3 is connected to WT2 and WT3 after optimal DSR. Moreover, the power flowing through each branch is regulated, and the overloading issue of some of branches is removed after optimal operation of system, as can be seen in Fig. 4.15.

As can be seen in Fig. 4.16, every CHS needs different value of G2V and V2G services at each hour of the day. In addition, the majority of V2G and G2V services are requested at valley period and peak period, respectively. Moreover, the drivers' social class changes the optimal demand/power of G2V/V2G services.

Table 4.3 Problem simulation results before (Sc. 1) and after (Sc. 2) optimal operation of system

Outputs	Sc.	Low-income drivers	Moderate-income drivers	High-income drivers
Operation cost ($/day)	Sc. 1	18,759		
	Sc. 2	11,806	14,036	16,117
Energy loss (MWh/day)	Sc. 1	367.90		
	Sc. 2	191.27	229.13	256.28
ENS (MWh/day)	Sc. 1	0.3499		
	Sc. 2	0.2932	0.3040	0.3057

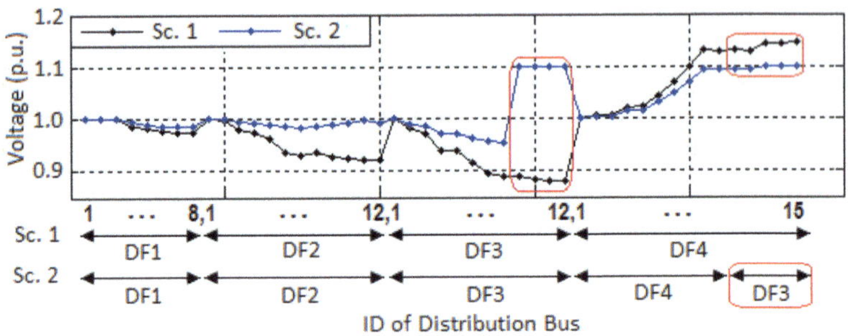

Fig. 4.14 Voltage profile of buses (p.u.) at 21st hour of the day (peak demand) before (Sc. 1) and after optimal operation of system (Sc. 2)

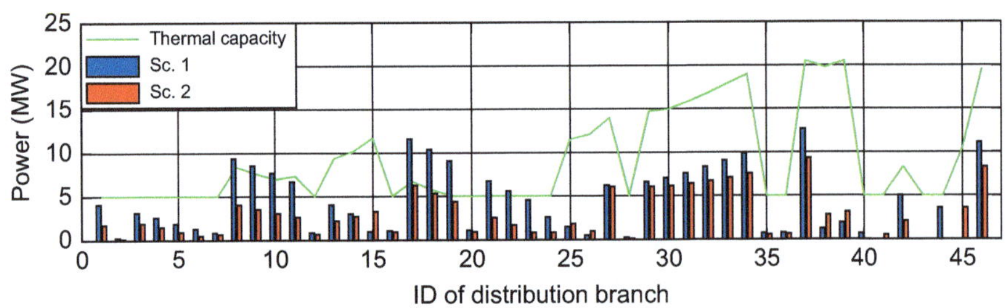

Fig. 4.15 Powers (MW) flowing through the branches and feeders' switches at 21st hour of the day (peak demand) before (Sc. 1) and after optimal operation of system (Sc. 2)

As can be seen in Fig. 4.17, each CHS suggests different scheme of incentive to the drivers. In addition, every CHS offers variable values of incentive throughout the day. Moreover, the drivers' social class affects the values of incentive that the low-income and high-income drivers need the least and the most incentives, respectively.

Figure 4.18 exhibits the optimal hourly status of feeders' switches. The indicated red-colored squares around some of the statuses demonstrate the difference between the results of low-income and high-income drivers to prove that the drivers' social

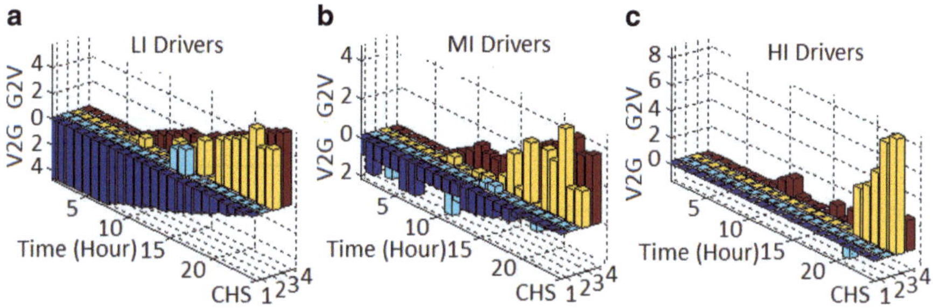

Fig. 4.16 Optimal hourly value of G2V or V2G at each CHS for each social class of drivers (Sc. 2)

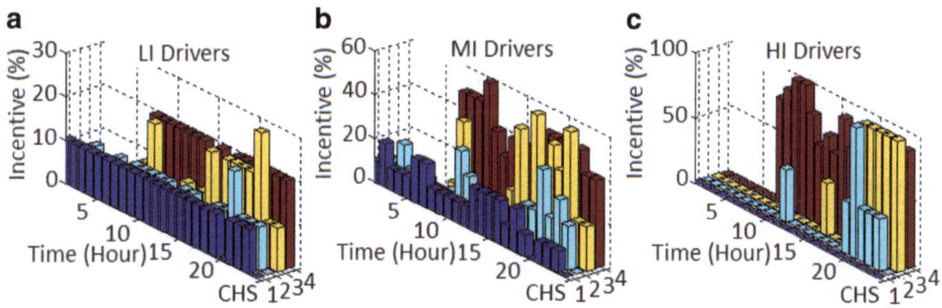

Fig. 4.17 Optimal hourly value of incentive at each CHS considering different social classes of drivers (Sc. 2)

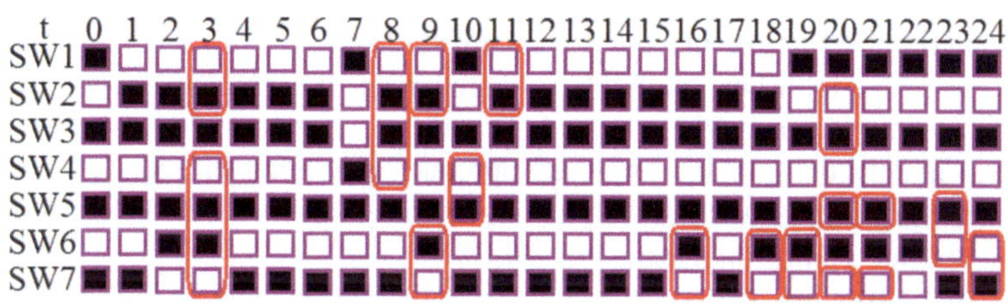

Fig. 4.18 Optimal hourly status of feeders' switches (Sc. 2) considering low-income drivers

class can influence the optimal status of switches and consequently the optimal configuration of system. As can be seen in Fig. 4.19, each ESS needs to change its mode (charging, discharging, or idle) with a specific pattern throughout the day.

4.4.2.3 Scenario 3

In this scenario, the robustness of proposed approach with respect to the prediction errors is studied. Herein, the problem is simulated with/without application of MPC considering different values of prediction error (10%, 20%, and 40%). In other

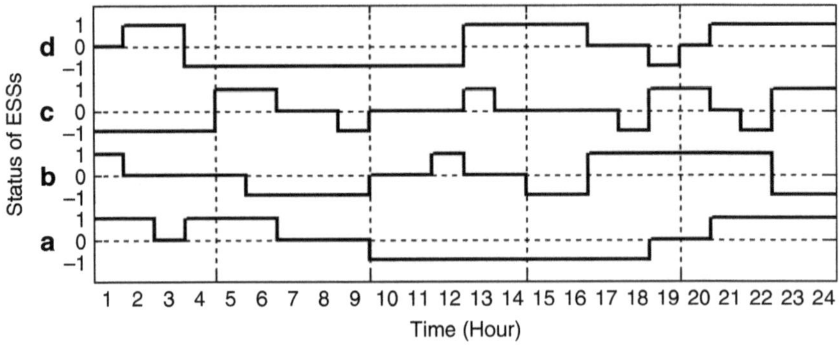

Fig. 4.19 Optimal hourly status of ESSs including charging, discharging, and idle statuses (Sc. 2)

Table 4.4 Simulation results of problem with/without application of MPC for different values of prediction error (Sc. 2–3)

–	With MPC (Sc. 2)	Without MPC (Sc. 3)		
Prediction error (%)	10, 20, 40	10	20	40
Operation cost ($/day)	11,806	11,894	12,355	14,705
Detriment ($/day)	–	88	549	2899

words, it is assumed that the predictions are not accurate, and there is an error in the value of uncertain states of problem (power of wind turbines and SOC of PEV fleet). The pattern of prediction errors includes successively underestimating and overestimating the value of uncertain states over the optimization time horizon. In this scenario, just the low-income drivers are studied.

In this scenario, when the MPC is disregarded, the problem is stochastically optimized once for the whole period. However, when the MPC is applied, the problem is solved using the proposed approach presented in Sect. 4.2. In other words, the problem is stochastically optimized at every time step (hour) of the day for the updated optimization time horizon.

As can be seen in Table 4.4, the application of MPC guarantees the robustness of problem outputs with respect to different values of prediction error, since the same results of Table 4.3 are achieved. As can be seen, the operation cost of system increases, and a detriment is imposed to the system operator when the MPC is not applied. In addition, the value of detriment grows when the prediction errors increase.

Figure 4.20 illustrates the optimal hourly status of feeders' switches and the optimal hourly mode of ESSs without application of MPC for 40% prediction error. The red-colored squares show the wrong status of switches (due to ignoring the application of MPC) and consequently the wrong configuration of system. The upward and downward arrows in Fig. 4.21 indicate the wrong modes of ESS because of neglecting the application of MPC.

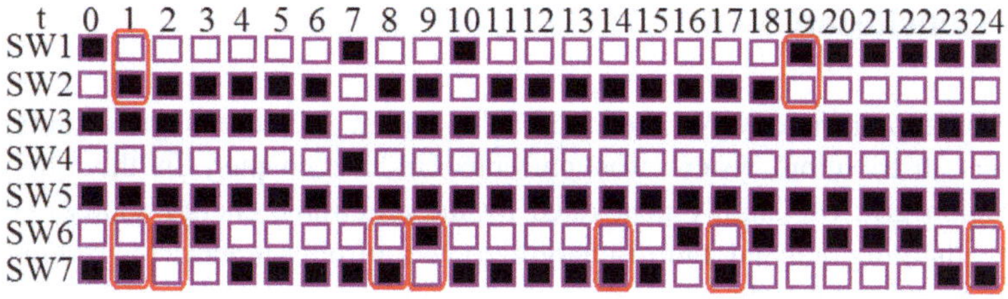

Fig. 4.20 Optimal hourly status of feeders' switches without application of MPC for 40% prediction error (Sc. 3)

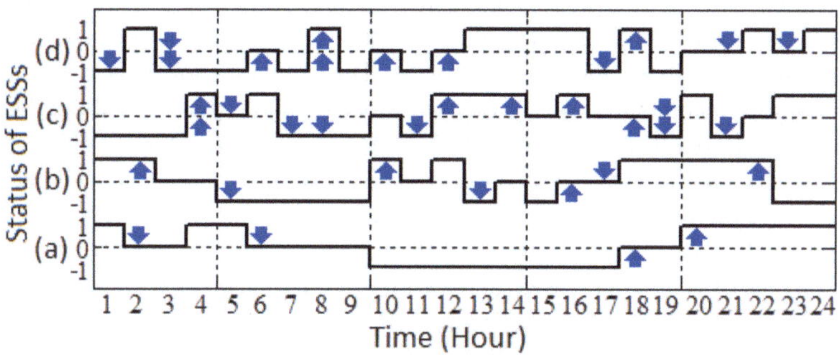

Fig. 4.21 Optimal hourly mode of ESSs without application of MPC for 40% prediction error (Sc. 3)

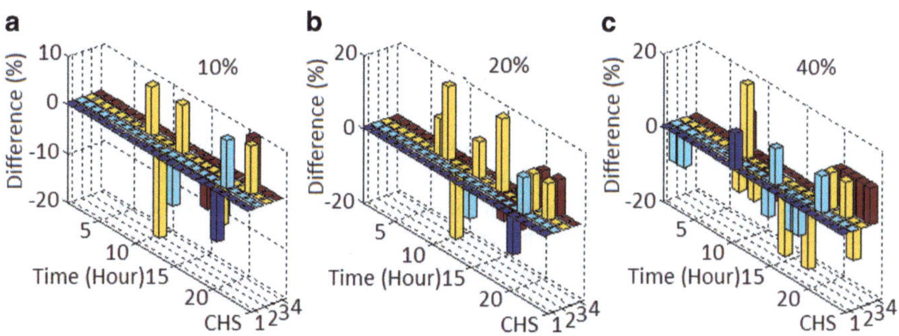

Fig. 4.22 Error in the hourly value of incentive of each CHS due to neglecting the application of MPC considering different values of prediction error (Sc. 3)

Figure 4.22 exhibits the error in hourly value of incentive at each CHS for different values of prediction error while disregarding the application of MPC. As can be seen, the existence of prediction error results in the wrong incentive scheme, and the wrong values increase as the prediction error grows.

4.5 Conclusion

A robust approach to solve the optimal operation problem of a reconfigurable electrical distribution system was presented in this chapter. In this study, the distribution grid was highly penetrated by the renewable energy sources (RESs), and the charging demand and discharging power of plug-in electric vehicle (PEV) fleet were optimally managed. Herein, the drivers' behavior was modelled based on the social and geographical aspects.

It was proven that the drivers' social class can affect the optimal mode of energy storage systems (ESS), optimal scheme of fleet management (FM), and optimal configuration of system. It was seen that the low-income and high-income drivers are the most and the least beneficial drivers for the system operator from the economic and reliability points of view, respectively. This phenomenon was related to the interest of low-income drivers and the reluctance of wealthy drivers about the value of incentive offered to them to provide the grid-to-vehicle (G2V) or vehicle-to-grid (V2G) services at the suggested charging stations (CHS) in the proposed periods.

It was observed that the application of model predictive control (MPC) makes the problem results robust with respect to different values of prediction error. In fact, the operation cost of system increased, and the misleading outcomes were achieved when the MPC was not applied. In other words, disregarding the application of proposed approach led to the suboptimal results and consequently unsatisfactory outcomes.

Nomenclature

Problem parameters	
β	Value of incentives including the discount on charging fee (for the G2V service provision) and the drivers' profit (as the percentage of electricity price for the V2G service provision)
ξ	Responsiveness probability of a driver (%)
μ	Mean value of hourly Gaussian distribution function
σ	Standard deviation of hourly Gaussian distribution function
τ, n^{τ}	Index of time (hour) of optimization time horizon and the length of optimization time horizon
π^{E}	Electricity price ($/MWh)
π^{ENS}	Damage ($/MWh) imposed to the customers because of a contingency like occurring a fault in a branch
π^{SW}	Feeder's switching cost ($)
π^{ESS}	ESS's switching cost ($)
λ	Failure rate of a branch of a feeder (failure/year)
φ^{SW}	Binary integer that indicates the status of a feeder's switch (open or close)
$\varphi^{G2V-V2G}$	Binary integer that indicates the status of vehicles to provide G2V or V2G service
φ^{ESS}	Binary integer that indicates the status of an ESS (charging, discharging, or idle)

(continued)

$\gamma_0^{CHS}, \gamma_1^{CHS}, \gamma_2^{CHS}, \gamma_3^{CHS}$	Binary integers used to show the binary value of incentive
b, B, N^B	Index, set, and total number of buses of system
br, BR, N^{BR}	Index, set, and total number of branches of system
chs, CHS, N^{CHS}	Index, set, and total number of charging stations of system
$Cost^{LOSS}$	Energy loss cost ($)
$Cost^{ENS}$	Energy not supplied cost ($)
$Cost^{DSR}$	Distribution system reconfiguration cost ($)
$Cost^{FM}$	Fleet management (FM) cost ($)
$Cost^{ESS}$	Energy storage system operation cost ($)
DOD^{ESS}	Depth of discharge defined for an energy storage system (%)
DOD^{PEVs}	Depth of discharge defined for a PEV (%)
e	Index of PEV
ess, ESS, N^{ESS}	Index, set, and total number of energy storage systems of system
$EENS$	Expected energy not supplied (MWh)
f, F, N^F	Index, set, and total number of feeders of system
f^{SOC}	Gaussian distribution function of SOC of PEV fleet
$f^{Weibull}$	Weibull distribution function
$\lvert I^{p.u.} \rvert$	Magnitude of current flowing through a branch (p.u.)
J	Cost-to-go function ($)
J^{Stoch}	Stochastic cost-to-go function ($)
k, η	Shape and scale parameters of a Weibull distribution function
LI, MI, HI	Indices of low-income, moderate-income, and high-income drivers
LNS^{FL-DSR}	Value of load not supplied during a fault locating and DSR intervals (MW)
LNS^{FR}	Value of load not supplied during a fault repairing period (MW)
MVA^{BASE}	Base MVA defined for the entire system (MVA)
$\lvert MVA \rvert$	Apparent power flowing through of a branch of a feeder (MVA)
$\overline{\lvert MVA \rvert}$	Thermal capacity constraint (power loading limit) of a branch of a feeder (MVA)
$\tilde{N}^{PEVs}$	Expected number of responsive drivers
$N^{PEVs,\ TOTAL}$	Total number of PEVs in the area
OF	Objective function of problem ($)
p	Occurrence probability of a scenario (%)
P^{LOSS}	Active power loss of a branch of a feeder (MW)
$\lvert P^{ESS} \rvert, \overline{P^{ESS}}$	Magnitude of power (MW) and maximum allowable power (MW) of an ESS
P^{PEV}	Power of a PEV (kW)
$\overline{P^{PEV}}$	Maximum allowable power of a PEV (kW)
P^{G2V}	Power of a PEV during providing G2V service (kW)
P^{V2G}	Power of a PEV during providing V2G service (kW)
$R^{p.u.}$	Resistance of a branch of a feeder in (p.u.)
s, S^{RWM}	Index of scenario and set of scenarios chosen by roulette wheel mechanism
S^{SOC}, S^W, S	Set of scenarios of SOC of PEV fleet, set of scenarios of wind speed (or power), and set of all the scenarios, respectively

(continued)

SC	Index of social class of drivers
SOC^{ESS}	State of charge of an ESS (%)
SOC^{PEVs}	State of charge of PEV fleet (%)
sw, SW, N^{SW}	Index, set, and total number of switches
t, T	Index of time (hour) and the number of hours of a day (24)
T^{FL-DSR}	Failure locating and DSR duration (hour)
T^{FR}	Failure repairing duration (hour)
U	Utility function ($)
$\|V\|, \underline{\|V\|}, \overline{\|V\|}$	Magnitude of voltage (p.u.) and minimum and maximum allowable voltage magnitudes of a bus (p.u.)
z	Discount factor ($0 \leq z < 1$)
IGA parameters	
θ^{M}	Mutation probability of a gene (%)
$\Delta Fitness$	Value of difference between the average fitness of chromosomes of new and old generations

Abbreviations

CHS	Charging station
DOD	Depth of discharge
DSR	Distribution system reconfiguration
EENS	Expected energy not supplied
ENS	Energy not supplied
ESS	Energy storage system
FM	Fleet management
G2V	Grid-to-vehicle
IGA	Improved genetic algorithm
LI, MI, HI	Low-income, moderate-income, high-income
MPC	Model predictive control
PEV	Plug-in electric vehicle
RES	Renewable energy source
RWM	Roulette wheel mechanism
SOC	State of charge
V2G	Vehicle-to-grid
WT	Wind turbine

References

1. R.E. Brown, *Electric Power Distribution Reliability* (Marcel Dekker, New York, 2002)
2. R. Billinton, R.N. Allan, *Reliability Evaluation of Power Systems*, 2nd edn. (Plenum, New York, 1996)
3. Monthly Plug-in Sales Scorecard. [Online]. Available: http://insideevs.com/monthly-plug-in-sales-scorecard. Accessed in Jan 2016

4. Electric Vehicles Initiative (EVI), Clean Energy Ministerial. [Online]. Available: http://cleanenergyministerial.org/Our-Work/Initiatives/Electric-Vehicles. Accessed on Jan 2016
5. [Online]. Available: https://www.greentechmedia.com/articles/read/everyone-is-revising-electric-vehicle-forecasts-upward#gs.eAwHRWdc. Accessed on Jan 2019
6. [Online]. Available: https://www.mckinsey.com/industries/automotive-and-assembly/our-insights/the-global-electric-vehicle-market-is-amped-up-and-on-the-rise. Accessed on Jan 2019
7. F.Y. Melhem, O. Grunder, Z. Hammoudan, N. Moubayed, Energy management in electrical smart grid environment using robust optimization algorithm. IEEE Trans. Ind. Appl. **54**(3), 2714–2726 (2018)
8. P.Y. Kong, A distributed management scheme for energy storage in a smart grid with communication impairments. IEEE Trans. Ind. Inf. **14**(4), 1392–1402 (2018)
9. M.R. Dorostkar-Ghamsari, M. Fotuhi-Firuzabad, M. Lehtonen, A. Safdarian, Value of distribution network reconfiguration in presence of renewable energy resources. IEEE Trans. Power Syst. **31**, 1879–1888 (2016)
10. N.G.A. Hemdana, B. Deppec, M. Pielked, M. Kurrata, T. Schmedesd, E. Wieben, Optimal reconfiguration of radial MV networks with load profiles in the presence of renewable energy based decentralized generation. Electr. Power Syst. Res. **116**, 355–366 (2014)
11. P. Meneses de Quevedo, J. Contreras, M.J. Rider, J. Allahdadian, Contingency assessment and network reconfiguration in distribution grids including wind power and energy storage. IEEE Trans. Sustain. Energy **6**, 1524–1533 (2015)
12. H. Haghighat, B. Zeng, Distribution system reconfiguration under uncertain load and renewable generation. IEEE Trans. Power Syst. **31**(4), 2666–2675 (2016)
13. J. Liu, H. Chiang, Maximizing available delivery capability of unbalanced distribution networks for high penetration of distributed generators. IEEE Trans. Power Del. **32**(3), 1196 (2017)
14. M.A. Rostami, A. Kavousi-Fard, T. Niknam, Expected cost minimization of smart grids with plug-in hybrid electric vehicles using optimal distribution feeder reconfiguration. IEEE Trans. Ind. Inf. **11**(2), 388–397 (2015)
15. A. Kavousi-Fard, T. Niknam, M. Fotuhi-Firuzabad, Stochastic reconfiguration and optimal coordination of V2G plug-in electric vehicles considering correlated wind power generation. IEEE Trans. Sustain. Energy **6**(3), 822–830 (2015)
16. M. Rahmani-Andebili, M. Fotuhi Firuzabad, An adaptive approach for PEVs charging management and reconfiguration of electrical distribution system penetrated by renewables. IEEE Trans. Ind. Inf. **14**(5), 2001–2010 (2018)
17. M. Rahmani-Andebili, G.K. Venayagamoorthy, SmartPark placement and operation for improving system reliability and market participation. Electr. Pow. Syst. Res. **123**(6), 21–30 (2015)
18. J.B. Rawlings, D.Q. Mayne, *Model Predictive Control: Theory and Design* (Nob Hill Publishing, LLC, Madison, 2009). [Online]. Available: http://jbrwww.che.wisc.edu/home/jbraw/mpc/electronic-book.pdf
19. M. Rahmani-Andebili, H. Shen, Energy scheduling for a smart home applying stochastic model predictive control, in *Proceedings of 25th Conference on Computer Communication and Networks*, Waikoloa, 1–4 Aug 2016
20. M. Rahmani-Andebili, Chapter 9: Cooperative distributed energy scheduling in microgrids, in *Electric Distribution Network Management and Control*, (Springer, 2018), pp. 235–254. Chapter DOI: https://doi.org/10.1007/978-981-10-7001-3_9. Book DOI: https://doi.org/10.1007/978-981-10-7001-3
21. M. Rahmani-Andebili, Chapter 11: Optimal incentive plans for plug-in electric vehicles, in *Electric Distribution Network Planning*, (Springer, 2018), pp. 299–320. Chapter DOI: https://doi.org/10.1007/978-981-10-7056-3_11. Book DOI: https://doi.org/10.1007/978-981-10-7056-3
22. A. Papoulis, S.U. Pillai, *Probability, Random Variables, and Stochastic Processes*, 4th edn. (McGraw-Hill, Boston, 2002). ISBN 0-07-366011-6

23. IEEE guide for electric power distribution reliability indices. IEEE Std. 1366–2003 (2004)
24. P. Wright, On minimum spanning trees and determinants. Math. Mag. **73**, 21–28 (2000)
25. M. Rahmani-Andebili, Dynamic and adaptive reconfiguration of electrical distribution system including renewables applying stochastic model predictive control. IET Gener. Transm. Distrib. **11**(16), 3912–3921 (2017)
26. D. Newbery, The economics of electric vehicles, in *EPRG and Imperial College London, E&E Seminar*, Cambridge, Jan 2013. [Online]. Available: http://www.eprg.group.cam.ac.uk/wp-content/uploads/2013/01/EEJan13_EconomicsEVs.p

Chapter 5
Optimal Operation of a Plug-In Electric Vehicle Parking Lot in the Energy Market Considering the Technical, Social, and Geographical Aspects

5.1 Introduction

Since electric vehicles and plug-in electric vehicles (PEVs) can be charged by the electricity generated by renewable energy sources, the internal combustion engine vehicles are being replaced by them to mitigate the energy security and environmental issues [1]. The PEV utilization has a rapid growth due to the battery technology advancement [2, 3]. Figure 5.1 displays the lithium-ion battery price ($/kWh) for the period of 2010–2016 [4]. As can be seen, this type of battery which is utilized in the electric vehicles has a descending price trend.

Figure 5.2 shows the time to sell one million electric vehicles and the cumulative worldwide sales of electric vehicles [5]. As can be seen, the interval is shortening as time goes on. In other words, the utilization of electric vehicles is increasing. In addition, Fig. 5.3 illustrates a prediction about the sales of oil-fueled cars and electric vehicles in the near future [6]. As can be noticed, the electric vehicle sales (including PEV sales) will surpass the gasoline and diesel cars by 2038.

A single PEV has a negligible impact on the electric power system; however, the uncontrolled charging/discharging of a large number of PEVs has a potential to negatively affect the power system's performance. In other words, the uncoordinated charging/discharging of PEVs may cause an overloading in the generation system, a congestion in the transmission and distribution lines, and even spikes in the energy market prices.

The PEVs are parked at the parking lots in the idle status for more than 90% during a typical day [7]. Therefore, the parking lot has access to a large size of electric capacity and power. In this regard, a PEV parking lot owner can play an effective role to mitigate the abovementioned issues, since the parking lot owner can motivate the PEV drivers, by offering a variety of incentives to them, to let him/her manage and coordinate their vehicles' charging time considering the drivers' expectation to have a full charge at the departure time. By implementing this plan, the

M. Rahmani-Andebili, *Planning and Operation of Plug-In Electric Vehicles*,
https://doi.org/10.1007/978-3-030-18022-5_5

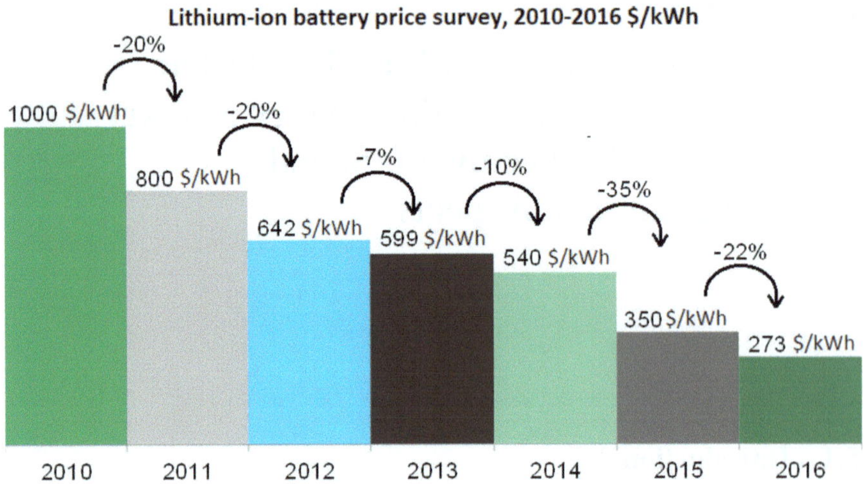

Fig. 5.1 Lithium-ion battery price ($/kWh) for the period of 2010–2016 [4]

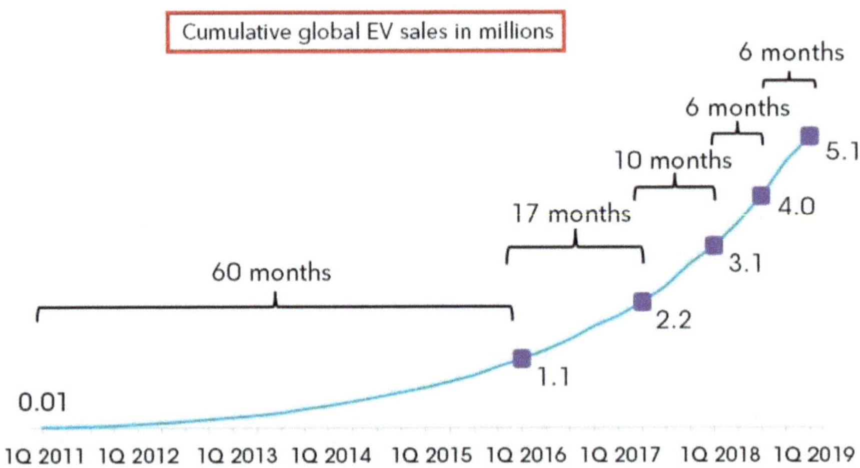

Fig. 5.2 The time to sell one million electric vehicles (EVs) and the cumulative worldwide sales of electric vehicles (millions) [5]. (Bloomberg NEF)

parking lot owner can take part in different power markets and provide profit for himself/herself, for the PEV drivers, and even for the electric power system.

The participation of parking lot in different power markets has been studied in [8–12]. In [8, 9], the PEV parking lot participates in the ancillary services including frequency regulation and spinning reserve market, respectively. In [10–12], the PEV parking lot participates in a day-ahead energy market. In [12], in addition to the drivers' behavior modelling, the capability of parking lot for the energy transaction and the battery life loss have been modelled.

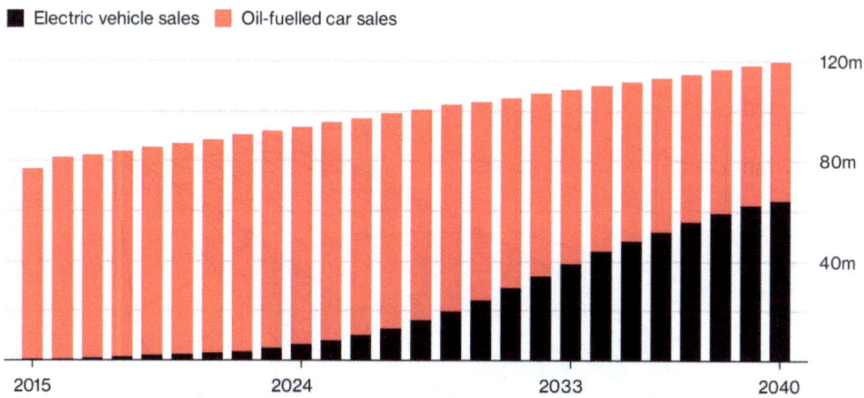

Fig. 5.3 The sales of electric vehicles and oil-fueled cars [6]

In this chapter, the problem of optimal operation of parking lot in the energy market is studied. Herein, the driving routes are modelled considering the minimum and maximum traffic speed limits and the real latitude and longitude of area around Marina City vertical parking lot in Chicago, IL 60654. In addition, the drivers' behavior is modelled considering the social and geographical factors. Furthermore, different PEV types, multiple PEV penetration levels, and various social classes of drivers are investigated in the operation problem. In this chapter, the problem is formulated as a mixed integer linear programming (MILP) problem.

In the following, in Sect. 5.2, the geographical, social, and technical aspects of problem are modelled. The problem formulation and the optimization technique are presented in Sects. 5.3 and 5.4, respectively. Section 5.5 is concerned with the problem simulation, and finally Sect. 5.6 concludes the chapter.

5.2 Modelling the Geographical, Social, and Technical Aspects of Problem

5.2.1 Drivers' Behavioral Models

In this study, the value of discount on charging fee of PEVs is defined as the value of incentive, and the drivers' behavior is modelled based their reaction with respect to that [13–15], their socioeconomic status (low-income, moderate-income, high-income), and their distance from the parking lot. Herein, it is logically assumed that the low-income and high-income drivers are the most and the least responsive drivers, since they are more and less likely to be interested in the proposed incentive to park their vehicles in the parking lot during the suggested period. Thus, a power function and an exponential function are applied to model their behavior with respect to value of incentive and their distance from the parking lot, respectively. In

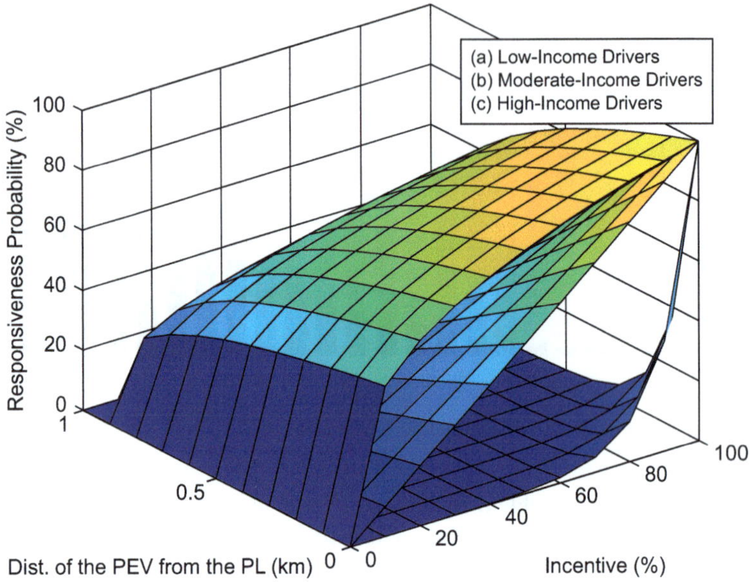

Fig. 5.4 Responsiveness probability of a driver with different social classes with respect to the value of incentive and his/her distance from the parking lot for the ranges of $0 \leq \beta \leq 100$ percent and $0 \leq \alpha \leq 1000$ meter

Table 5.1 Responsiveness probability of a low-income, moderate-income, and high-income driver with respect to the value of incentive (β) and his/her distance from the parking lot (α)

Social class	Driver's responsiveness probability (percent)
Low-income (LI)	$\xi_{LI,\alpha,\beta} = 100 \times \left(1 - \frac{\alpha}{1000}\right)^{0.3} \times \left(\frac{\beta}{100}\right)^{0.3}$ (5.1)
Moderate-income (MI)	$\xi_{MI,\alpha,\beta} = 100 \times \left(1 - \frac{\alpha}{1000}\right) \times \left(\frac{\beta}{100}\right)$ (5.2)
High-income (HI)	$\xi_{HI,\alpha,\beta} = 100 \times exp\left(-10 \times \frac{\alpha}{1000}\right) \times exp\left(10 \times \left(\frac{\beta}{100} - 1\right)\right)$ (5.3)

addition, a linear function is applied to model the behavior of moderate-income drivers with respect to value of above mentioned parameters.

The responsiveness probability functions of a low-income, moderate-income, and high-income driver with respect to the value of incentive (β) and his/her distance (α) from the parking lot for the ranges of $0 \leq \beta \leq 100$ percent and $0 \leq \alpha \leq 1000$ meter are graphically shown in Fig. 5.4 and presented in Table 5.1. In Table 5.1, *LI*, *MI*, and *HI* are the indices of low-income, moderate-income, and high-income drivers, respectively.

As can be seen in Fig. 5.4, these three models approximately include all the possibilities for the driver's reaction, since the whole three-dimensional space is rationally covered by them. Herein, some logics can be realized from the graphs and Table 5.1, presented as follow:

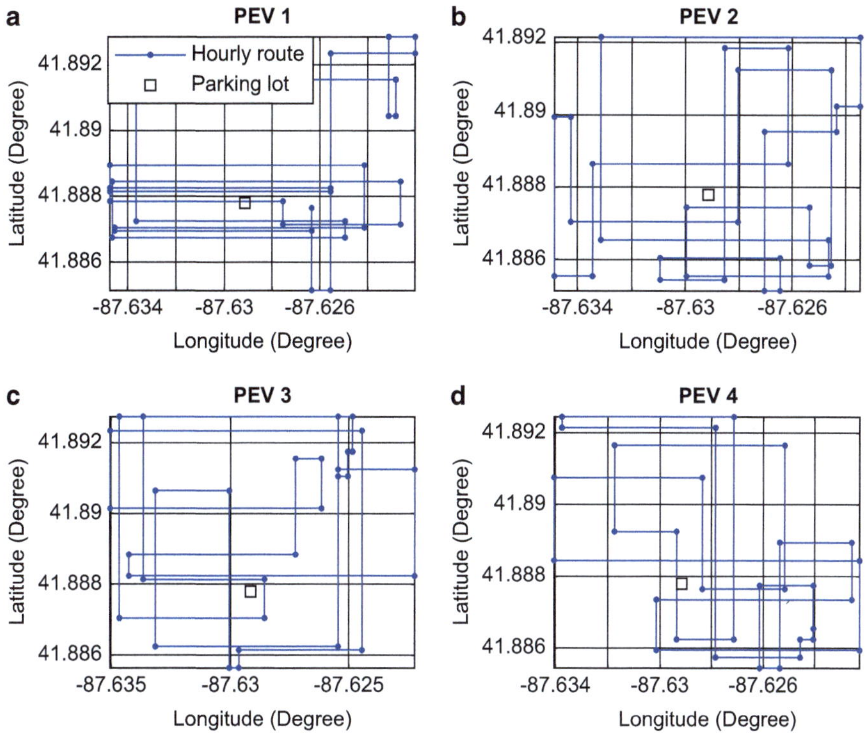

Fig. 5.5 Hourly driving routes concerned with PEVs 1–4 in the typical day considering the minimum and maximum traffic speed limits and the real latitude and longitude of area around Marina City vertical parking lot, Chicago, IL 60654, USA

- The driver, with any incentive, will not park his/her vehicle at the parking lot if the parking lot is farther than the maximum specified distance ($\alpha > 1000$ meters).
- The driver, with any distance from the parking lot, will not park his/her vehicle at the parking lot if there is no incentive ($\beta = 0\%$).
- The driver, with the nearest distance to the parking lot ($\alpha \leq 100$ meters), will definitely park his/her vehicle at the parking lot if the incentive maximum ($\beta = 100\%$).
- The possibility of occurrence of other cases is proportional to the probability value presented in Table 5.1 and graphically illustrated in Fig. 5.4.

5.2.2 Driving Routes

Figures 5.5, 5.6, and 5.7 show the location of parking lot and the hourly driving routes of ten PEVs in the typical day. These driving patterns have been randomly generated by a computer while considering the minimum and maximum traffic speed

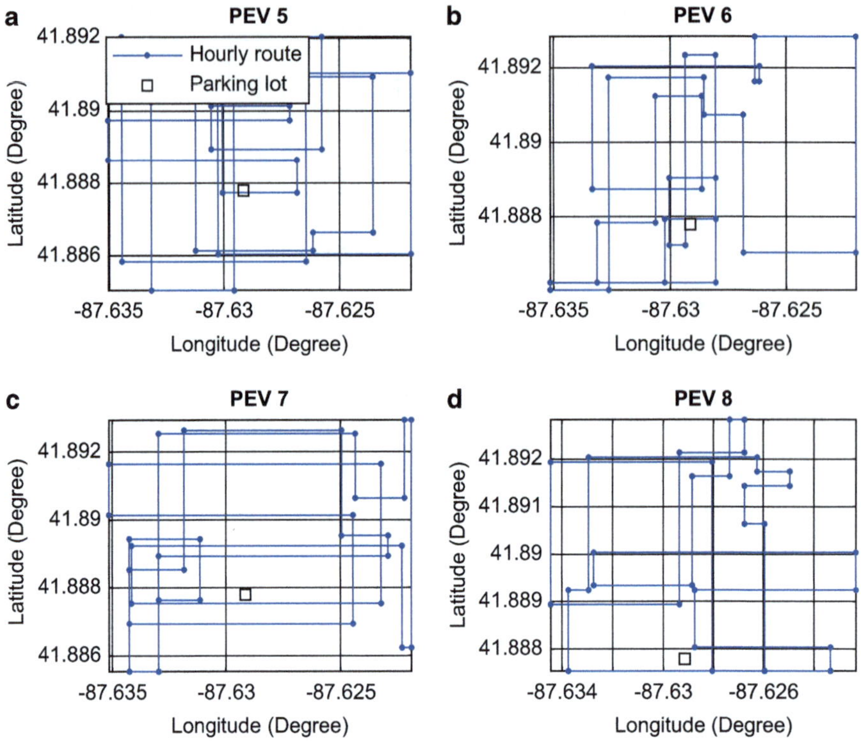

Fig. 5.6 Hourly driving routes concerned with PEVs 5–8 in the typical day considering the minimum and maximum traffic speed limits and the real latitude and longitude of area around Marina City vertical parking lot, Chicago, IL 60654, USA

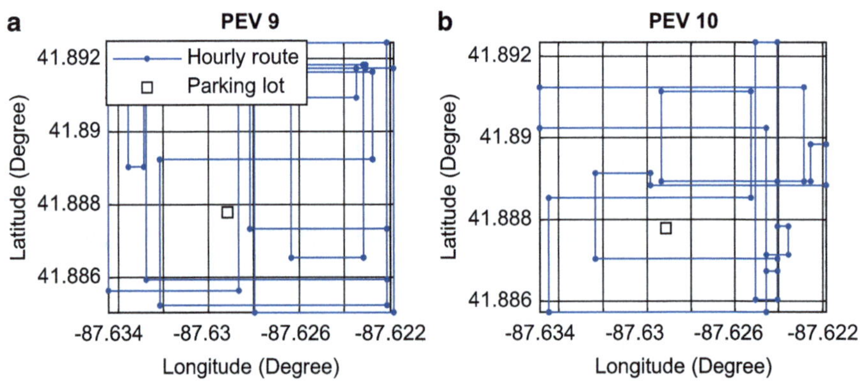

Fig. 5.7 Hourly driving routes concerned with PEVs 9–10 in the typical day considering the minimum and maximum traffic speed limits and the real latitude and longitude of area around Marina City vertical parking lot, Chicago, IL 60654, USA

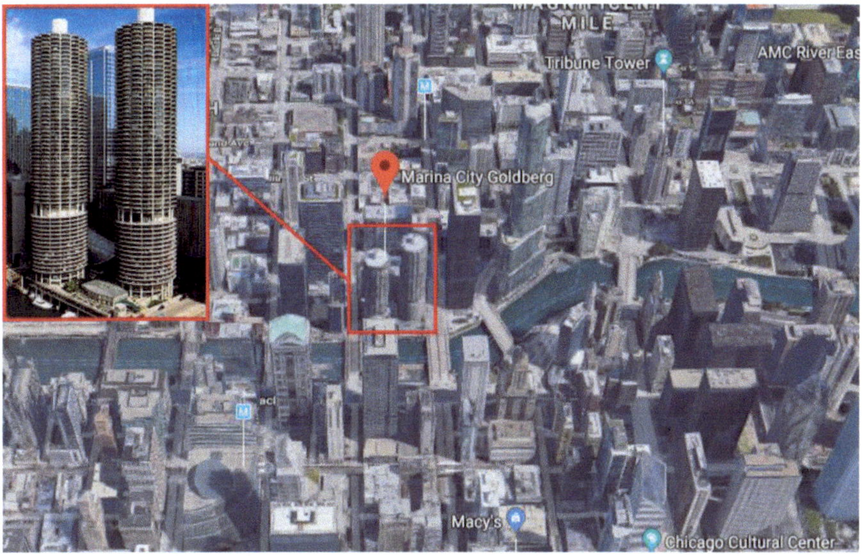

Fig. 5.8 The location of Marina City vertical parking lot in Chicago, IL 60654, USA [17]. The real latitude and longitude of parking lot (left-top, right-top, left-bottom, and right-bottom) and four corners of area are (−87.629139, 41.887791), (−87.636783, 41.893932), (−87.618072, 41.892932), (−87.636783, 41.885039), and (−87.618072, 41.885039), respectively

limits [16] and the real latitude and longitude of area around Marina City vertical parking lot in Chicago, IL 60654, USA. As can be seen in Figs. 5.5, 5.6, and 5.7, at each hour of the day, each PEV has different distance from the parking lot. Therefore, even the drivers related to an identical social class will have dissimilar reactions with respect to an equal value of incentive.

Figure 5.8 illustrates the Marina City vertical parking lot located in Chicago, IL 60654, USA. The real latitude and longitude of parking lot (left-top, right-top, left-bottom, and right-bottom) and four corners of area are (−87.629139, 41.887791), (−87.636783, 41.893932), (−87.618072, 41.892932), (−87.636783, 41.885039), and (−87.618072, 41.885039), respectively.

5.2.3 State of Charge of PEVs

In this study, a descending trend is considered for the state of charge (SOC) of PEVs in the typical day, since the PEVs lose their charge as they travel. In addition, a relatively larger standard deviation is assumed for the SOC of PEV fleet during the time period of 8–22. Based these assumptions, the hourly mean and one-sigma uncertainty band of SOC of PEV fleet in the operation time horizon (1 day) is shown in Fig. 5.9. Moreover, Fig. 5.10 illustrates the hourly mean and one-sigma uncertainty band of SOC of PEV fleet for the extended operation time horizon

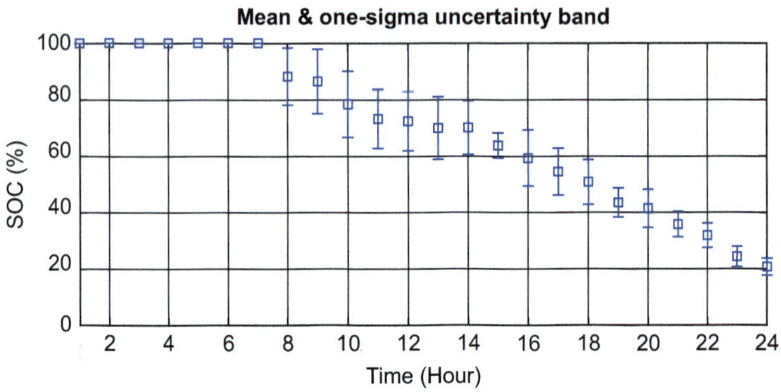

Fig. 5.9 The hourly mean and one-sigma uncertainty band of SOC of PEV fleet (percent) in the operation time horizon (1 day)

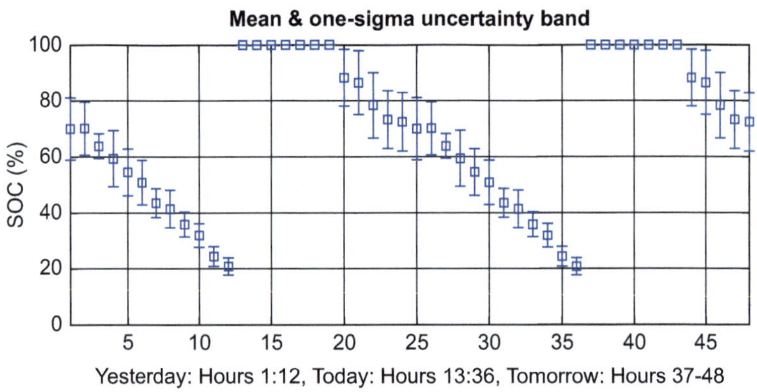

Fig. 5.10 The hourly mean and one-sigma uncertainty band of SOC of PEV fleet (percent) in the extended operation time horizon (2 days)

(2 days). In Fig. 5.10, the interval of 13–36 indicates the operation time horizon, and the periods 1–12 and 37–48 show the parts of the day before and after operation time horizon, respectively. As can be seen, the same descending trend has been considered for the extended operation time horizon.

5.2.4 *Arrival and Departure Time of PEVs*

In this study, the arrival time and departure time of each PEV are randomly generated by a computer for the extended time horizon. Herein, during the data generation process, it is considered that the departure time must be after the arrival

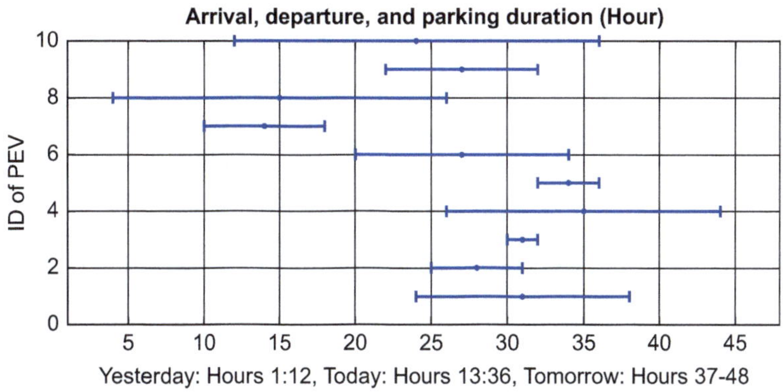

Fig. 5.11 The arrival time, departure time, and parking duration (hour) of responsive PEVs in the extended operation time horizon

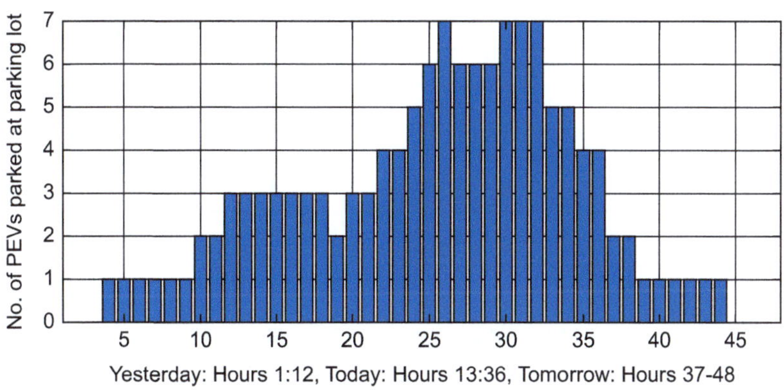

Fig. 5.12 The total number of PEVs parked at the parking lot at each hour of extended operation time horizon

time. In addition, the PEVs may arrive to the parking lot prior the operation time horizon or leave it after the operation time horizon. Figure 5.11 shows the arrival time, the departure time, and the parking duration (in hour) of responsive PEVs during the extended operation time horizon.

Based on the parking duration of PEVs, the hourly number of PEVs, which are present at the parking lot, can be determined. Figure 5.12 shows the total number of PEVs parked at the parking lot at each hour of extended operation time horizon. In addition, as can be seen in Fig. 5.13, the total power of PEVs (MW) parked at the parking lot is calculated considering the rated power of PEVs. Herein, the type of PEVs is Tesla Model S that has the rated power of 225 kW [18]. As can be seen, the parking lot has different hourly power during the extended operation time horizon.

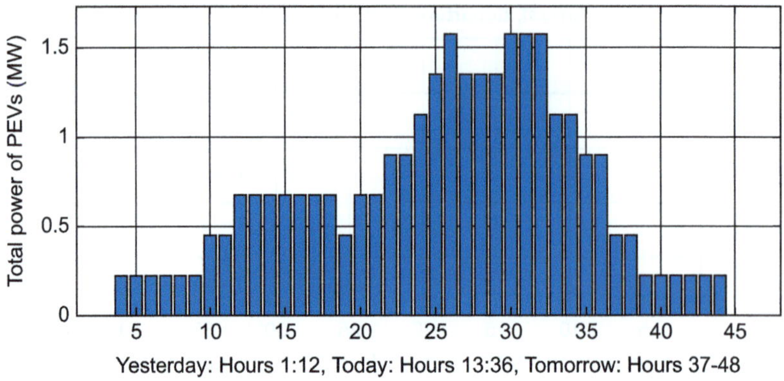

Fig. 5.13 The total power of PEVs (MW) parked at the parking lot at each hour of extended operation time horizon

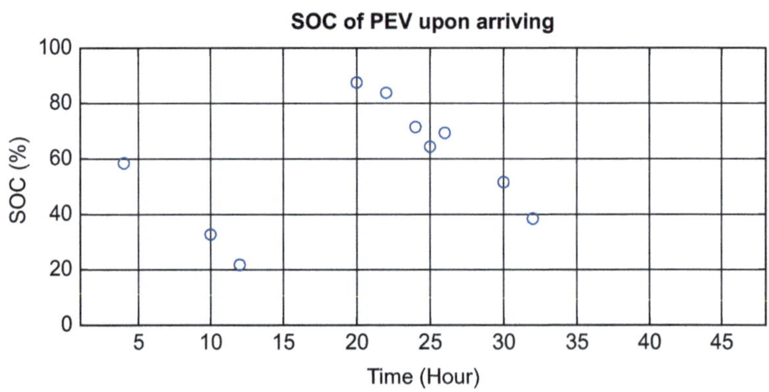

Fig. 5.14 The SOC of each responsive PEV (percent) upon its arrival to the parking lot during the extended operation time horizon

Using the mean and one-sigma uncertainty band of SOC of PEV fleet (presented in Fig. 5.10), the SOC of each responsive PEV, upon its arrival to the parking lot, is randomly determined, as is seen in Fig. 5.14.

Considering the hourly number of PEVs parked at the parking lot, the SOC of PEVs upon their arrival to the parking lot, and the rated capacity of battery of Tesla Model S (60 kWh) [18], the hourly energy saved in the PEVs (MWh) is calculated and shown in Fig. 5.15. Furthermore, the inaccessible energy, available vehicle-to-grid (V2G) energy, and available grid-to-vehicle (G2V) energy of PEVs (MWh) are calculated for the extended operation time horizon using the allowable depth of discharge (DOD) and rated power and capacity of PEVs [12], as is shown in Fig. 5.16. As can be noticed, both V2G and G2V services are available at each hour of operation time horizon; however, it is up to the parking lot owner to decide to optimally choose one or none of services (i.e., idle status).

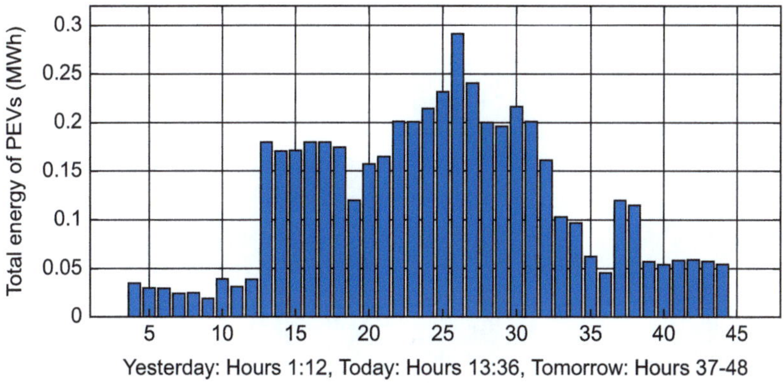

Fig. 5.15 The total energy saved in the PEVs (MWh) parked at the parking lot at each hour of extended operation time horizon

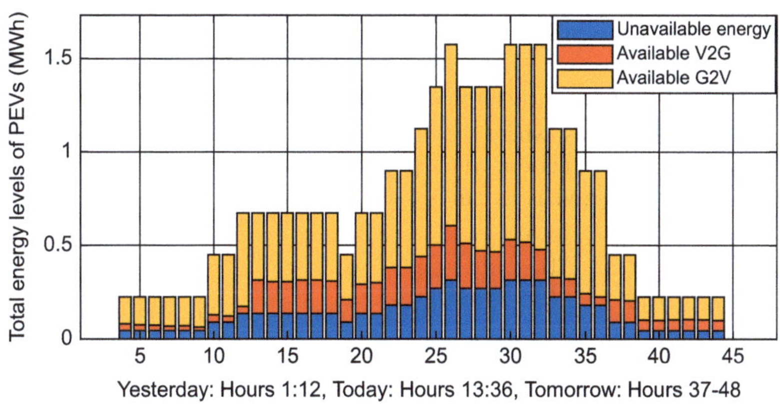

Fig. 5.16 The hourly inaccessible energy, available V2G energy, and available G2V energy of PEVs (MWh) parked at the parking lot during the extended operation time horizon

5.2.5 Equipping the Parking Lot with Renewables

In this study, it is assumed that the vertical parking lot is equipped by the renewable energy sources including wind turbines and photovoltaic panels. By the application of renewables, the vertical parking lot can supply part of its demand and even sell back its extra electrical energy to the power market through the electrical distribution grid. The Bahrain World Trade Center, Manama, Bahrain [19], is an example of application of wind turbines in a skyscraper, as can be seen in Fig. 5.17. In fact, the Bahrain World Trade Center that includes three 225 kW wind turbines is the first structure that integrates the commercial-scale wind turbines into a building [19].

Fig. 5.17 The Bahrain World Trade Center including three 225 kW wind turbines on the bridges spanning the Twin Towers, Manama, Bahrain [19]

The output power of a wind turbine (P^W in MW) has a nonlinear relationship with the wind speed (v) in m/s, where v^{ci}, v^r, v^{co}, and $\overline{P^W}$ are the cut-in wind speed (m/s), rated wind speed (m/s), cut-out wind speed (m/s), and rated power of wind turbine (MW) [22, 23], respectively, as can be seen in Eq. (5.4).

$$P^W(t) = \begin{cases} 0 & v(t) & < v^{ci}, v(t) > v^{co} \\ \overline{P^W} \times \dfrac{v^f(t) - v^{ci}}{v^r - v^{ci}} & & v^{ci} \leq v(t) \leq v^r \\ \overline{P^W} & & v^r \leq v(t) \leq v^{co} \end{cases}, \quad \forall t \in \{1, \ldots, T\} \tag{5.4}$$

Moreover, the vertical parking lot can be supplied by solar energy using photovoltaic panels installed on its walls and roof. As can be seen in Fig. 5.18, the CIS Solar Tower in Manchester, England, is an example of application of solar energy in a skyscraper that has been equipped by 390 kW photovoltaic panels [20]. A numerical study regarding a solar parking lot has been presented in [21].

Moreover, as can be seen in Eq. (5.5), the output power of photovoltaic panels (MW) is a nonlinear function of solar irradiance (ρ) in W/m^2, where ρ^{st}, ρ^c, and $\overline{P^{PV}}$ are the solar irradiance in the standard environment (1000 W/m^2), certain solar irradiation point set as 150 W/m^2, and rated power of photovoltaic panels (MW), respectively [22, 23]. In addition, the rated power of photovoltaic panels is calculated using Eq. (5.6), where A^{PV} and η^{PV} are the total surface area of parking lot that has been covered by the photovoltaic panels (m^2) and the efficiency of photovoltaic panels (percent), respectively.

Fig. 5.18 CIS Solar Tower including 390 kW photovoltaic panels installed on its walls, Manchester, England [20]

$$P^{PV}(t) = \begin{cases} \overline{P^{PV}} \times \dfrac{(\rho(t))^2}{\rho^{st} \times \rho^c} & \rho(t) \leq \rho^c \\ \overline{P^{PV}} \times \dfrac{\rho(t)}{\rho^{st}} & \rho(t) > \rho^c \end{cases}, \forall t \in \{1, \ldots, T\} \tag{5.5}$$

$$\overline{P^{PV}} = A^{PV} \times \frac{\eta^{PV}}{100} \tag{5.6}$$

5.3 Problem Formulation

5.3.1 *Objective Function*

In this part of chapter, the objective function concerned with the optimal operation problem of PEV parking lot in the energy market is presented. As can be seen in Eq. (5.7), the objective function includes several income and cost terms, that is, the cost of discount on charging fee of PEVs as the incentive cost ($Cost^{INC}$ in \$), the cost of purchasing electrical energy form the power market ($Cost^{BUY}$ in \$), the income of selling electrical energy to the market ($Income^{SELL}$ in \$), the degradation cost of

PEVs' batteries due to the V2G service provision ($Cost^{V2G}$ in \$), and the degradation cost of PEVs' batteries due to the G2V service provision ($Cost^{G2V}$ in \$). Herein, it is assumed that the parking lot owner is responsible to pay the PEVs' batteries degradation cost to the drivers. In Eq. (5.7), t is the index of time (hour) and T indicates the operation time horizon (24 hours).

$$OF = max\left\{-Cost^{INC} + \sum_{t=1}^{T} - Cost^{BUY}(t) + Income^{SELL}(t) - Cost^{V2G}(t) - Cost^{G2V}(t)\right\} \tag{5.7}$$

5.3.1.1 Incentive Cost

The incentive cost term (\$) is presented in Eq. (5.8). As can be seen, it depends on the responsiveness probability of drivers (ξ in percent), social class of drivers (SC), distance of PEVs from the parking lot (α in meter), discount on charging fee of PEVs (β in percent), rated capacity of each responsive PEV's battery (Cap^{PEV} in kW), initial SOC of each responsive PEV ($SOC^{PEV}(0)$ in percent), and minimum price of energy market (π^{M-MIN} in \$/MWh). Herein, the PEVs parked at the parking lot will be charged based on the minimum electricity price. Moreover, the drivers that their vehicles stay in the parking lot for at least 8 hours will receive the incentive.

In Eq. (5.8), e and N indicate the index of PEV and total number of PEVs in the area, and LI, MI, and HI stand for low-income, moderate-income, and high-income, respectively.

$$\begin{aligned} Cost^{INC} = & \sum_{e=1}^{N} \frac{\xi_{e,SC,\alpha,\beta}}{100} \times \frac{Cap_e^{PEV}}{1000} \times \left(1 - \frac{SOC_e^{PEV}(0)}{100}\right) \\ & \times \pi^{M-MIN} \times \frac{\beta}{100}, \quad \forall SC \in \{LI, MI, HI\}, \forall \alpha \in \{0, 100, \ldots, 1000\}, \\ & \forall \beta \in \{0, 10, \ldots, 100\} \end{aligned} \tag{5.8}$$

5.3.1.2 Electricity Purchasing Cost

The cost term concerned with the electricity purchasing cost (\$) is presented in (5.9). As can be seen, it is the function of hourly market price and purchasing power (P^{BUY} in MW). Herein, Y^{BUY}, as a binary integer variable, indicates the purchasing status of parking lot, where it is equal to one if the parking lot purchases the electrical energy from the power market.

$$Cost^{BUY}(t) = \pi^{M}(t) \times P^{BUY}(t) \times Y^{BUY}(t), \ \ \forall t \in \{1, \ldots, T\} \tag{5.9}$$

5.3.1.3 Electricity Selling Income

The income term ($) that depends on the market price and selling power (P^{SELL} in MW) is presented in Eq. (5.10). The binary integer variable Y^{SELL} represents the selling status of parking lot, where it is equal to one if the parking lot sells back the electricity to the market.

$$Income^{SELL}(t) = \pi^{M}(t) \times P^{SELL}(t) \times Y^{SELL}(t), \ \ \forall t \in \{1, \ldots, T\} \tag{5.10}$$

5.3.1.4 PEVs' Batteries Degradation Cost

The hourly degradation cost terms of PEVs' batteries due to the V2G and G2V service provision ($) are presented in Eqs. (5.11) and (5.12), respectively. Herein, C^{V2G} and C^{G2V} indicate the battery degradation cost of one PEV ($) due to the provision of V2G and G2V services for 1 hour ($), respectively. In addition, Y^{V2G} and Y^{G2V}, as the binary integer variables, indicate the status of PEV to provide the V2G and G2V services, respectively, where they are nonzero if the PEV is discharged and charged, respectively.

$$Cost^{V2G}(t) = \sum_{e=1}^{N} \frac{\xi_{e,SC,\alpha,\beta}}{100} \times C^{V2G} \times Y_{\mathrm{e}}^{V2G}(t), \ \ \forall t \in \{1, \ldots, T\} \tag{5.11}$$

$$Cost^{G2V}(t) = \sum_{e=1}^{N} \frac{\xi_{e,SC,\alpha,\beta}}{100} \times C^{G2V} \times Y_{\mathrm{e}}^{G2V}(t), \ \ \forall t \in \{1, \ldots, T\} \tag{5.12}$$

5.3.2 Equality Constraints

In the following, the equality constraints of problem related to the optimal operation of parking lot in the energy market are presented and described.

5.3.2.1 Demand-Generation Balance Constraint

The demand-generation balance constraint is considered in the problem at each hour of the operation period to guarantee the equality of power generation and load demand, as can be seen in Eq. (5.13). Herein, a positive sign is considered for the power of wind turbine (P^{W} in MW), power of photovoltaic panels (P^{PV} in MW), power of V2G service of PEVs (P^{V2G} in MW), and energy purchasing power (P^{BUY} in MW), while a negative sign is assumed for the load demand (P^{D} in MW), power of G2V service of PEVs (P^{G2V} in MW), and energy selling power (P^{SELL} in MW). The V2G power and G2V power of PEVs (kW) are presented in Eqs. (5.14) and (5.15), respectively. As can be seen, they are defined as the expected value of power of V2G and G2V services, where they are proportional to the responsiveness probability of drivers.

$$\begin{aligned} &P^{BUY}(t) + P^{V2G}(t) - P^{SELL}(t) - P^{G2V}(t) + P^{PV}(t) + P^{W}(t) - P^{D}(t) \\ &\quad = 0, \ \ \forall t \in \{1, \ldots, T\} \end{aligned} \tag{5.13}$$

$$P^{V2G}(t) = \sum_{e=1}^{N} \frac{\xi_{e,SC,\alpha,\beta}}{100} \times \frac{p_e^{V2G}(t)}{1000} \tag{5.14}$$

$$P^{G2V}(t) = \sum_{e=1}^{N} \frac{\xi_{e,SC,\alpha,\beta}}{100} \times \frac{p_e^{G2V}(t)}{1000} \tag{5.15}$$

5.3.2.2 PEV's SOC Constraint

The hourly SOC of each responsive PEV (percent) is determined using Eq. (5.16), where p^{G2V}, Cap^{PEV}, and $Eff^{PEV}$are the power (kW) and capacity (kWh) of battery of PEV and the efficiency of PEV's battery (percent) to provide G2V or V2G service, respectively. Herein, n indicates the expected number of responsive PEVs, which is determined using Eq. (5.17).

$$\begin{aligned} \frac{SOC_e^{PEV}(t)}{100} = &\frac{SOC_e^{PEV}(t-1)}{100} + \frac{p_e^{G2V}(\mathrm{t})}{Cap_e^{PEV}} \times \frac{Eff_e^{PEV}}{100} \\ &- \frac{p_e^{V2G}(\mathrm{t})}{Cap_e^{PEV}} \times \frac{1}{Eff_e^{PEV}/100}, \quad \forall t \in \{1, \ldots, T\}, \ \ \forall e \in \{1, \ldots, n\} \end{aligned} \tag{5.16}$$

$$n = \sum_{e=1}^{N} \frac{\xi_{e,SC,\alpha,\beta}}{100} \tag{5.17}$$

5.3.3 *Inequality Constraints*

In this part, the inequality constraints concerned with the optimal operation of parking lot in the electricity market are presented.

5.3.3.1 Parking Lot's Maximum Power Constraint

To model the nominal power of equipment installed in the parking lot, a power limit ($\overline{P^{PL}}$) is considered on the hourly amount of energy that the parking lot purchases (P^{BUY}) from the market through the electrical distribution system, as can be seen in Eq. (5.18). Herein, the binary integer variable (Y^{BUY}) indicates the purchasing status of parking lot.

$$P^{BUY}(t) \leq \overline{P^{PL}} \times Y^{BUY}(t), \ \ \forall t \in \{1, \ldots, T\} \tag{5.18}$$

In addition, as can be seen in (5.19), there is also a power limit on the hourly amount of energy that the parking lot sells back (P^{SELL}) to the electricity market. The binary integer variable (Y^{SELL}) represents the selling status of parking lot.

$$P^{SELL}(t) \leq \overline{P^{PL}} \times Y^{SELL}(t), \ \ \forall t \in \{1, \ldots, T\} \tag{5.19}$$

5.3.3.2 Buying-Selling Constraint

This constraint is held to prevent the parking lot from simultaneously purchasing and selling back the electrical energy, which is a meaningless action. As can be seen in (5.20), by applying this constraint, just one of the binary integer variables (Y^{BUY} and Y^{SELL}) or none of them will be equal to one that implies the possibility of existence of idle status for the parking lot in some intervals of operation period.

$$Y^{BUY}(t) + Y^{SELL}(t) \leq 1, \ \ \forall t \in \{1, \ldots, T\} \tag{5.20}$$

5.3.3.3 V2G and G2V Power Constraint

The battery of a PEV has a rated power that restricts it from energy transaction beyond this limit. As can be seen in Eqs. (5.21) and (5.22), the V2G (discharging) and G2V (charging) powers must be equal or less than the nominal power of PEV. Herein, the binary integer variables (Y^{V2G} and Y^{G2V}) indicate the status of PEV to provide the V2G and G2V services, respectively.

$$p_e^{V2G}(t) \le \overline{P_e^{PEV}} \times Y_e^{V2G}(t),\ \ \forall t \in \{1,\ldots,T\},\ \ \forall e \in \{1,\ldots,n\} \tag{5.21}$$

$$p_e^{G2V}(t) \le \overline{P_e^{PEV}} \times Y_e^{G2V}(t),\ \ \forall t \in \{1,\ldots,T\}, \forall e \in \{1,\ldots,n\} \tag{5.22}$$

5.3.3.4 V2G-G2V Constraint

Similar to the buying-selling constraint of parking lot, a V2G-G2V constraint is applied to prevent each responsive PEV from simultaneously being charged (G2V service provision) and discharged (V2G service provision), as can be seen in Eq. (5.23). Therefore, by applying this constraint, the PEV might be in charging, discharging, or idle status at each hour of operation period.

$$Y_e^{V2G}(t) + Y_e^{G2V}(t) \le 1,\ \ \forall t \in \{1,\ldots,T\},\ \ \forall e \in \{1,\ldots,n\} \tag{5.23}$$

5.3.4 Lower and Upper Bounds of Variables

In this section, the lower and upper bounds of continuous and integer variables are presented. In Eqs. (5.24) and (5.25), the lower and upper power limits of parking lot during the energy purchasing and selling are presented, respectively. Moreover, the lower and upper power limits of V2G and G2V services of each responsive PEV are presented in Eqs. (5.26) and (5.27), respectively. As can be seen in Eq. (5.28), DOD^{PEV} (in percent) and 100% are considered as the lower and upper bounds of SOC of each responsive PEV. In addition, Y^{V2G}, Y^{G2V}, Y^{BUY}, and Y^{SELL}, as the binary integer variables, have the lower and upper bounds of 0 and 1, as can be seen in Eqs. (5.29, 5.30, 5.31, and 5.32), respectively.

$$0 \le P^{BUY}(t) \le \overline{P^{PL}},\ \ \forall t \in \{1,\ldots,T\} \tag{5.24}$$

$$0 \le P^{SELL}(t) \le \overline{P^{PL}},\ \ \forall t \in \{1,\ldots,T\} \tag{5.25}$$

$$0 \le p_e^{V2G}(t) \le \overline{P^{PEV}},\ \ \forall t \in \{1,\ldots,T\},\ \ \forall e \in \{1,\ldots,n\} \tag{5.26}$$

$$0 \le p_e^{G2V}(t) \le \overline{P^{PEV}},\ \ \forall t \in \{1,\ldots,T\},\ \ \forall e \in \{1,\ldots,n\} \tag{5.27}$$

$$DOD^{PEV} \le SOC_e^{PEV}(t) \le 100,\ \ \forall t \in \{1,\ldots,T\},\ \ \forall e \in \{1,\ldots,n\} \tag{5.28}$$

$$0 \le Y_e^{V2G}(t) \le 1,\ \ \forall t \in \{1,\ldots,T\},\ \ \forall e \in \{1,\ldots,n\} \tag{5.29}$$

$$0 \le Y_e^{G2V}(t) \le 1,\ \ \forall t \in \{1,\ldots,T\},\ \ \forall e \in \{1,\ldots,n\} \tag{5.30}$$

$$0 \leq Y^{BUY}(t) \leq 1, \ \ \forall t \in \{1, \ldots, T\} \tag{5.31}$$

$$0 \leq Y^{SELL}(t) \leq 1, \ \ \forall t \in \{1, \ldots, T\} \tag{5.32}$$

5.4 Mixed Integer Linear Programing as the Optimization Technique

As can be noticed from Sect. 5.3, the parking lot operation problem has been formulated as a MILP in this chapter. Thus, in this part, the MILP and its matrices and vectors are described.

In a MILP, the objective function ($[f]^T[X]$) has a linear form, where $[f]$ is a column vector of constants and $[X]$ is a column vector of unknown variables, as can be seen in Eq. (5.33) [24, 25]. Herein, $[X]$ includes both continuous and integer values. Additionally, MILP has some linear equality constraints, presented by $[A_{eq}]$ matrix and $[b_{eq}]$ vector, and some linear inequality constraints, presented by $[A]$ matrix and $[b]$vector, as can be seen in Eq. (5.33). Moreover, as is seen in Eq. (5.33), there are lower and upper bounds on the unknown variables, presented by $[l_b]$ and $[u_b]$ vectors. More descriptions about MILP can be found in [24, 25].

$$MILP: \begin{cases} min\left\{[f]^T[X]\right\} \\ [A][X] \leq [b] \\ [A_{eq}][X] = [b_{eq}] \\ [X] \leq [u_b] \\ [X] \geq [l_b] \end{cases} \tag{5.33}$$

MILP is normally a minimization algorithm. Therefore, in order to set the optimization technique as a maximization one, the objective function needs to change its sign, as can be seen in Eq. (5.34). In this chapter, a MILP is applied as an optimization algorithm to solve the optimal operation problem of parking lot in the energy market.

$$max\left\{[f]^T[X]\right\} = min\left\{-[f]^T[X]\right\} \tag{5.34}$$

Considering Eqs. (5.7)–(5.32), the dimension of unknown variables vector ($[X]$) and column vector of constants of objective function ($[f]$) will be $24(4 + 5n) \times 1$, as can be seen in Eq. (5.35). Herein, "24," "4," "5," and "n" indicate the number of hours of operation period (1 day), the total number of continuous and integer variables concerned with the parking lot, the total number of continuous and integer variables concerned with a responsive PEV, and the expected number of responsive PEVs in the parking lot.

In addition, the linear equality constraints matrix and vector ($[A_{eq}]$ and $[b_{eq}]$) have the dimensions of $24(1 + n) \times 24(4 + 5n)$ and $24(1 + n) \times 1$, respectively, as can be seen in Eq. (5.36). Herein, "1" and "1" in $(1 + 1n)$ (or $(1 + n)$) represent the single linear equality constraint concerned with the parking lot and the sole linear equality constraint concerned with a responsive PEV, respectively.

Moreover, as can be seen in Eq. (5.37), the linear inequality constraints matrix and vector ($[A]$ and $[b]$) have the dimensions of $24(3 + 3n) \times 24(4 + 5n)$ and $24(3 + 3n) \times 1$, respectively. Herein, "3" and "3" represent the number of linear inequality constraints concerned with the parking lot and the number of linear inequality constraints concerned with a responsive PEV, respectively.

Furthermore, the lower and upper bounds vectors ($[l_b]$ and $[u_b]$) have the same dimension of unknown variables vector, that is, $24(4 + 5n) \times 1$, as can be seen in Eq. (5.38).

$$[X] = [\cdot]_{24(4+5n)\times 1}, [f] = [\cdot]_{24(4+5n)\times 1} \tag{5.35}$$

$$[A_{eq}] = [\cdot]_{24(1+n)\times 24(4+5n)}, [b_{eq}] = [\cdot]_{24(1+n)\times 1} \tag{5.36}$$

$$[A] = [\cdot]_{24(3+3n)\times 24(4+5n)}, [b] = [\cdot]_{24(3+3n)\times 1} \tag{5.37}$$

$$[l_b] = [\cdot]_{24(4+5n)\times 1}, [u_b] = [\cdot]_{24(4+5n)\times 1} \tag{5.38}$$

5.5 Problem Simulation

5.5.1 Primary Data of Problem

In this chapter, Marina City vertical parking lot and the area around it are studied. The parking lot and the related area, located in Chicago, IL 60654, USA, are illustrated in Fig. 5.8. The typical hourly driving routes of ten PEVs considering the minimum and maximum traffic speed limits and the real latitude and longitude of area around Marina City vertical parking lot have been shown in Figs. 5.5, 5.6, and 5.7.

The technical specifications of different PEV types are presented in Table 5.2 [18]. As can be seen, the PEVs have unidentical rated powers and capacities. The DOD, considered for each PEV (DOD^{PEV}), is 20% in this chapter. The efficiency of

Table 5.2 The technical specifications of some PEV types [18]

Type of PEV	Tesla Model S	Citroën C-Zero	Volkswagen e-Up	Renault Kangoo Z. E.
Battery power (kW)	225	49	60	44
Battery capacity (kWh)	60	16	18.7	22

Table 5.3 Categorizing the PEV penetration levels based on the number of PEVs in the area and the driving routes

		PEV penetration level		
–		Low	Moderate	High
Number of PEVs in the area with different driving routes	Driving route 1	100	300	500
	⋮	⋮	⋮	⋮
	Driving route 10	100	300	500
Total number of PEVs in the area		1000	3000	5000

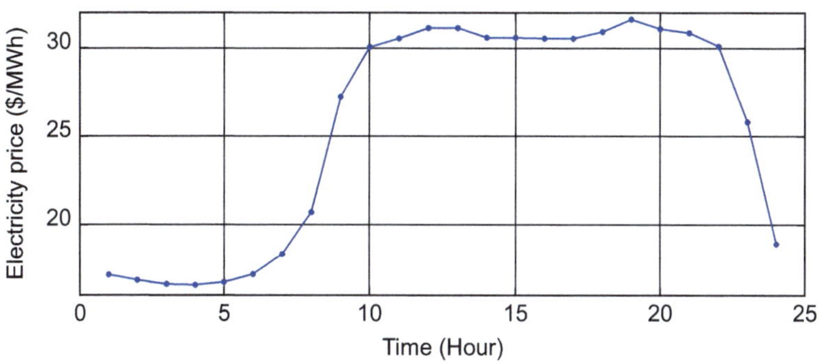

Fig. 5.19 The hourly electricity price of power market ($/MWh) [26]

a PEV's battery to provide G2V or V2G service is considered about 95%. The degradation cost of one PEV's battery due to the provision of V2G service (C^{G2V}) or G2V service (C^{V2G}) is about \$0.01/hour. The rated power of equipment installed in the parking lot ($\overline{P^{PL}}$) is 10 MW.

Table 5.3 presents the PEV penetration levels categorized based on the number of PEVs in the area and the driving routes. The arrival time, departure time, and parking duration (hour) of responsive PEVs are randomly determined using the proposed approach presented in Sect. 5.2.4. The hourly mean and one-sigma uncertainty band of SOC of PEV fleet during the operation period have been given in Fig. 5.9. The responsiveness probability of a driver with respect to the value of incentive (β) and his/her distance from the parking lot (α) for the ranges of $0 \leq \beta \leq 100$ percent and $0 \leq \alpha \leq 1000$ meter for each social class (low-income, moderate-income, and high-income) has been illustrated in Fig. 5.4.

Figure 5.19 shows the hourly electricity price of energy market [26]. The hourly wind speed (v) and solar irradiance in the typical day are illustrated in Fig. 5.20. Additionally, Fig. 5.21 displays the hourly load demand of parking lot, and the hourly powers of wind turbine and photovoltaic panels are calculated using Eqs. (5.4, 5.5, and 5.6). The load demand of parking lot is concerned with the car lift stock installed in the parking lot. The rated power of wind turbine ($\overline{P^{W}}$), cut-in wind speed (v^{ci}), rated wind speed (v^{r}), and cut-out wind speed (v^{co}) are about 8 MW,

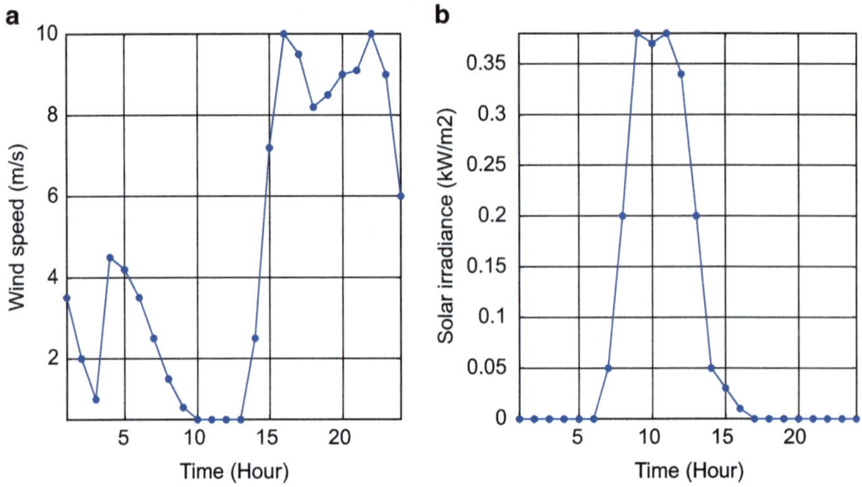

Fig. 5.20 (**a**) The wind speed (m/s) and (**b**) solar irradiance (kW/m^2) in the typical day

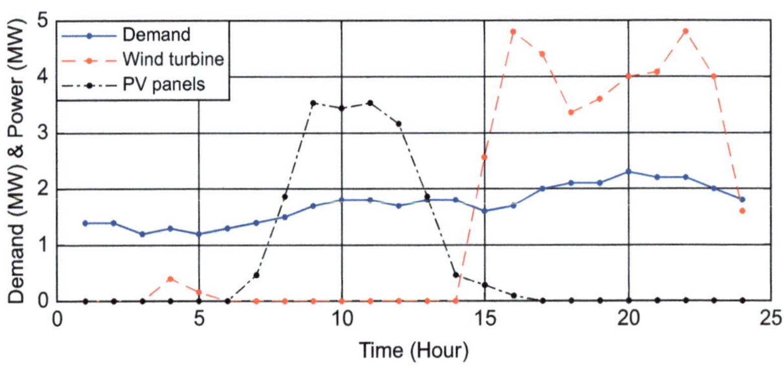

Fig. 5.21 The hourly load demand (MW) and powers (MW) of wind turbine and photovoltaic panels

4 m/s, 14 m/s, and 25 m/s, respectively. Furthermore, the solar irradiance in the standard environment (ρ^{st}), the certain solar irradiation point (ρ^{c}), the surface area of parking lot covered by the photovoltaic panels (A^{PV}), and the efficiency of photovoltaic panels (η^{PV}) are 1000 W/m^2, 150 W/m^2, 37200 m^2, and 25%, respectively.

Figure 5.22 illustrates the diagram of load and different energy sources connected to the PEV parking lot. As can be seen, the parking lot is supplied by the renewable energy sources such as wind turbine and photovoltaic panels, and it has the capability of bilateral energy transaction with the power market through the electrical distribution system and with the PEV fleet using the V2G and G2V services. Additionally, the parking lot has its own load demand needed to lift the PEVs.

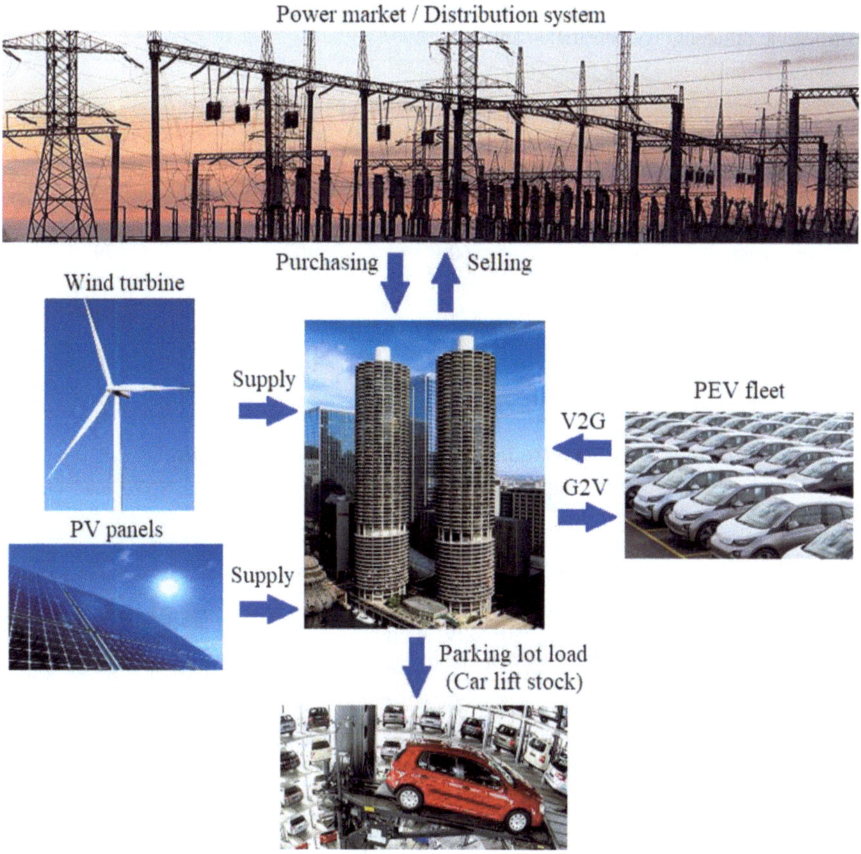

Fig. 5.22 The diagram of load and energy sources connected to the PEV parking lot

5.5.2 *Studying the Problem with Tesla Model S*

5.5.2.1 Low-Income Drivers

In this part of problem simulation, the type of PEVs and the social class of drivers are assumed Tesla Model S and low-income, respectively. Table 5.4 presents the problem simulation results for each PEV penetration level, that is, low, moderate, and high. As can be seen, the parking lot has initially an income about $558/day before solving the parking lot's optimal operation problem. This income is resulted from selling back the excess energy supplied by the renewable energy sources of parking lot.

However, after the optimal implementation of incentive plan and participation of parking lot in the energy market, parking lot's profit increases for any PEV penetration level. As can be seen, studying the case of high PEV penetration level results in the most profit for the parking lot ($1362/day). In addition, as can be observed, the

Table 5.4 The problem simulation results for Tesla Model S PEVs and low-income drivers considering different PEV penetration levels

Outputs	Before operation	After optimal operation of parking lot		
	Any penetration	Low penetration	Moderate penetration	High penetration
Optimal incentive (%)	0	30	20	20
Discount cost ($)	0	60	107	179
Income ($/day)	558	939	1255	1542
Maximum profit ($/day)	558	878	1147	1362
Enhancement ($/day)	–	319	588	803

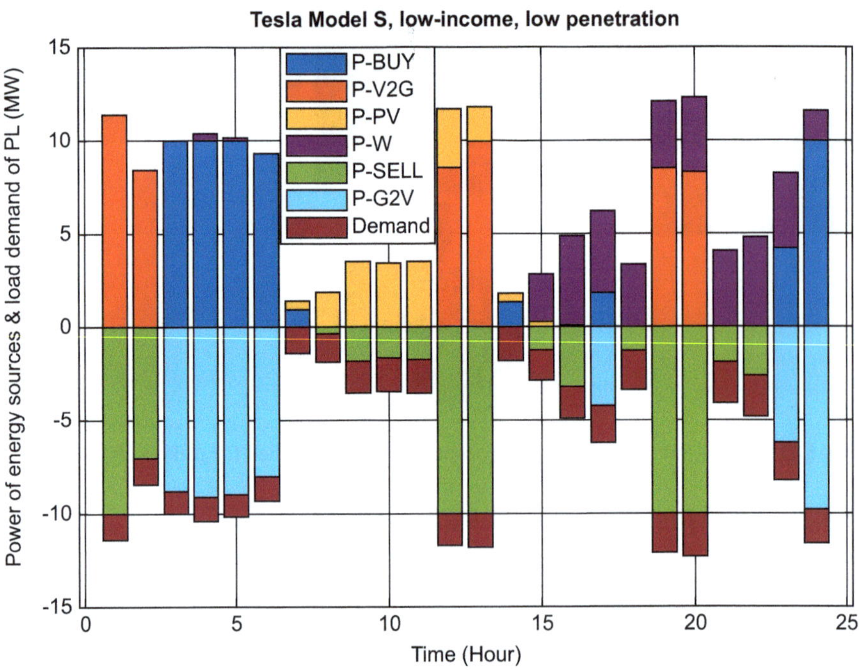

Fig. 5.23 The optimal hourly energy scheduling of different energy sources of parking lot considering Tesla Model S PEVs, low-income drivers, and low PEV penetration level

optimal value of incentive is different for the low PEV penetration level that proves the necessity of investigation of each penetration level.

Figures 5.23, 5.24, and 5.25 illustrate the optimal hourly energy scheduling of different energy sources of parking lot for Tesla Model S PEVs and low-income drivers considering different PEV penetration levels. These figures show the correctness of hourly demand-generation balance constraint, since the total energy generated by the energy sources of parking lot (positive values) is equal to the

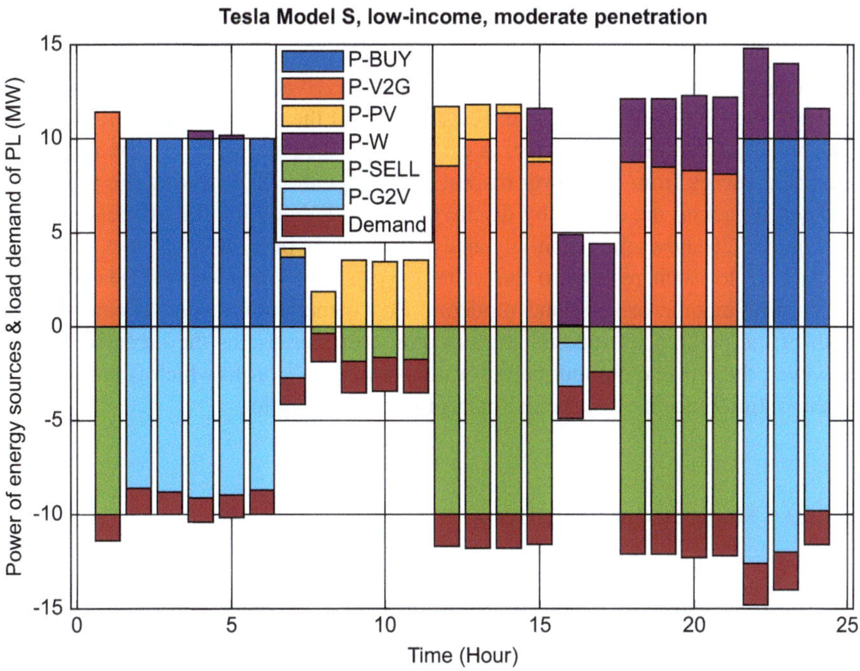

Fig. 5.24 The optimal hourly energy scheduling of different energy sources of parking lot considering Tesla Model S PEVs, low-income drivers, and moderate PEV penetration level

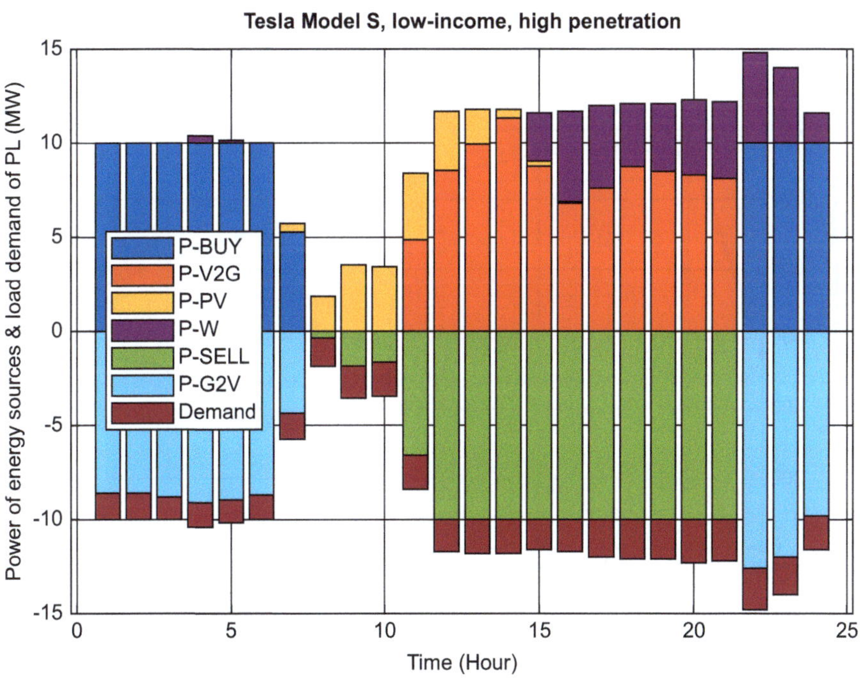

Fig. 5.25 The optimal hourly energy scheduling of different energy sources of parking lot considering Tesla Model S PEVs, low-income drivers, and high PEV penetration level

total energy consumed by the rest of them (negative values). In addition, as can be noticed, the PEV penetration level is able to affect the optimal operation of energy sources of parking lot. Moreover, as can be seen, the parking lot prefers to buy the electrical energy from the power market and charge the PEVs during the peak price and sell it back to the market by discharging the PEVs during the off-peak price.

Figure 5.26 shows the sensitivity analysis for the income, discount cost, and profit of parking lot with respect to the value of incentive considering Tesla Model S PEVs, low-income drivers, and moderate PEV penetration level. As can be seen, both of the parking lot income and discount cost increase as the incentive grows; however, there is one optimal point for the parking lot's profit which is proportional to the optimal incentive and maximum profit of parking lot.

5.5.2.2 Moderate-Income Drivers

In this section, similar to the previous case, the problem is simulated for different PEV penetration levels considering Tesla Model S as the type of PEVs; however, the social class of drivers is changed to moderate-income. Table 5.5 presents the problem simulation results. Like the previous case, the high PEV penetration level

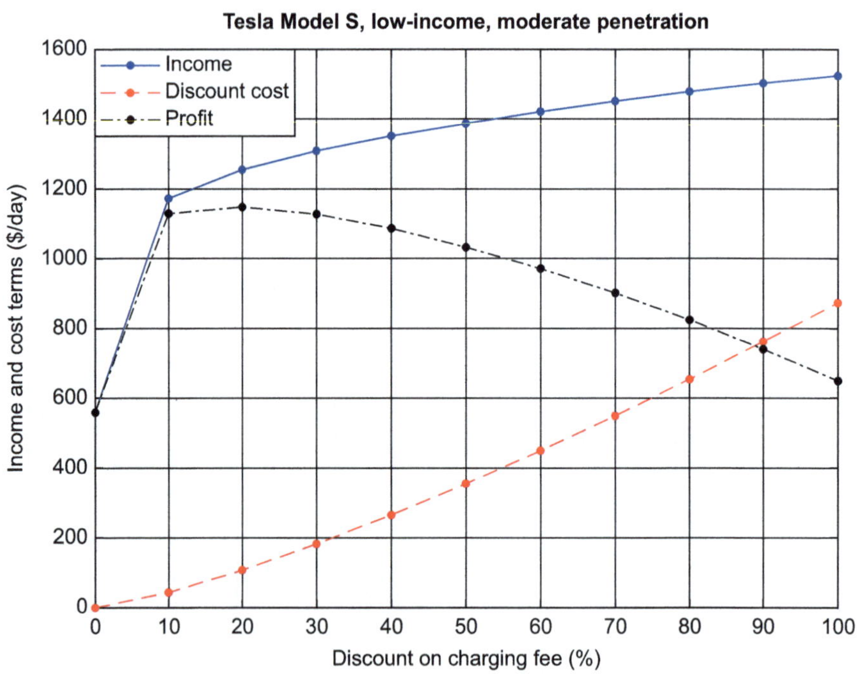

Fig. 5.26 The sensitivity analysis for the income, discount cost, and profit of parking lot with respect to the value of incentive considering Tesla Model S PEVs, low-income drivers, and moderate PEV penetration level

Table 5.5 The problem simulation results for Tesla Model S PEVs and moderate-income drivers considering different PEV penetration levels

	Before operation	After optimal operation of parking lot		
Outputs	Any penetration	Low penetration	Moderate penetration	High penetration
Optimal incentive (%)	0	70	40	40
Discount cost ($)	0	139	136	228
Income ($/day)	558	934	1088	1279
Maximum profit ($/day)	558	794	951	1051
Enhancement ($/day)	–	236	393	492

results in the most profit for the parking lot ($1051/day). Nevertheless, compared to the previous case, the optimal incentives change and increase, and the maximum profit of parking lot decreases for any PEV penetration level.

The optimal hourly energy scheduling of energy sources of parking lot for each PEV penetration level is shown in Figs. 5.27, 5.28, and 5.29. As can be seen, the demand-generation balance constraint is held at each hour of operation period.

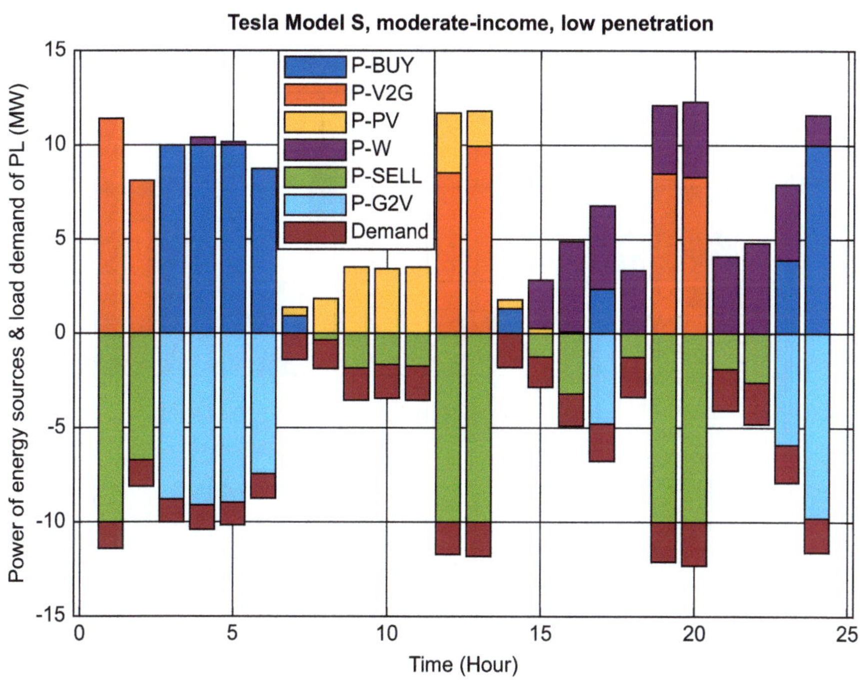

Fig. 5.27 The optimal hourly energy scheduling of different energy sources of parking lot considering Tesla Model S PEVs, moderate-income drivers, and low PEV penetration level

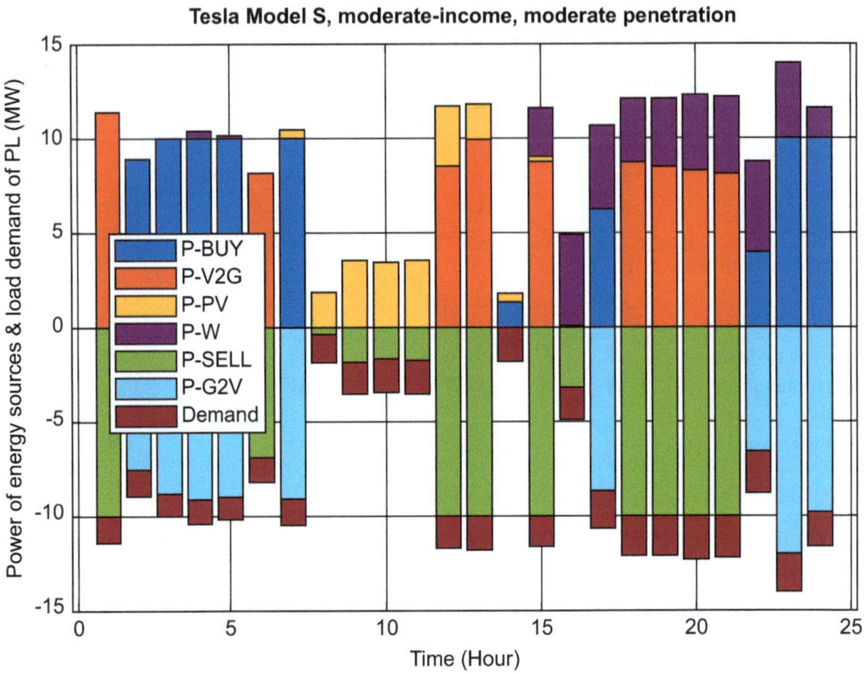

Fig. 5.28 The optimal hourly energy scheduling of different energy sources of parking lot considering Tesla Model S PEVs, moderate-income drivers, and moderate PEV penetration level

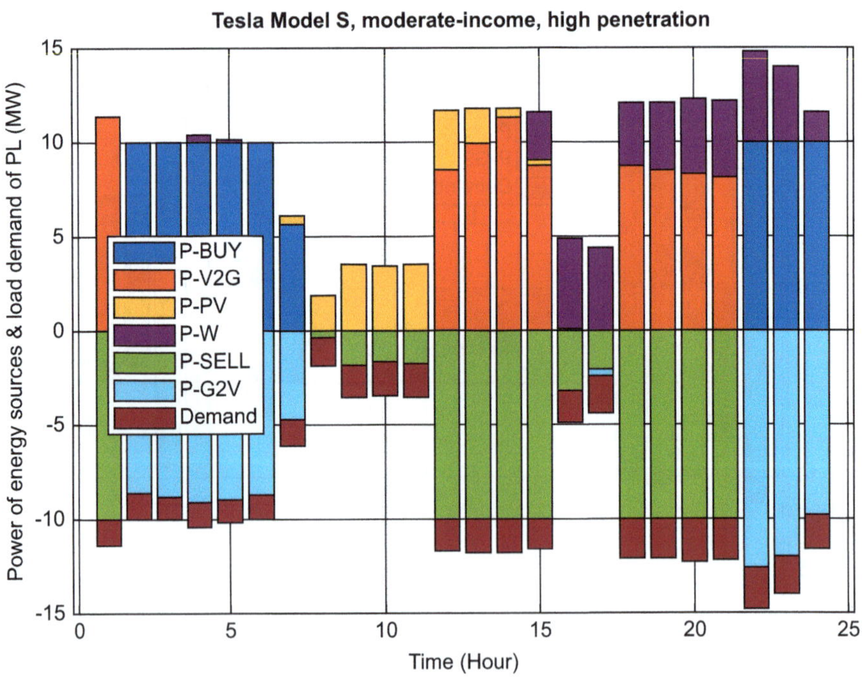

Fig. 5.29 The optimal hourly energy scheduling of different energy sources of parking lot considering Tesla Model S PEVs, moderate-income drivers, and high PEV penetration level

Herein, like the previous case, the electricity purchasing and G2V service provision occur during the peak electricity price, and the V2G service is provided, and the electrical energy is sold back to the market during the off-peak electricity price. However, by comparing these figures with Figs. 5.23, 5.24, and 5.25, it is noticed that changing the social class of drivers causes many changes in the optimal operation of energy sources of parking lot, especially in the cases of moderate and high PEV penetration levels.

The sensitivity analysis for the income, discount cost, and profit of parking lot with respect to the value of incentive considering Tesla Model S PEVs, moderate-income drivers, and moderate PEV penetration level is illustrated in Fig. 5.30. As can be seen, the parking lot income and discount cost have ascending trends with respect to the value of incentive; however, there is a maximum profit for the parking lot and an optimal incentive for the PEVs.

5.5.2.3 High-Income Drivers

In this part, like the previous cases, the type of PEVs is Tesla Model S, and the problem is simulated for different PEV penetration levels; nonetheless, it is assumed that the drivers are related to the high-income social class. Table 5.6 gives the

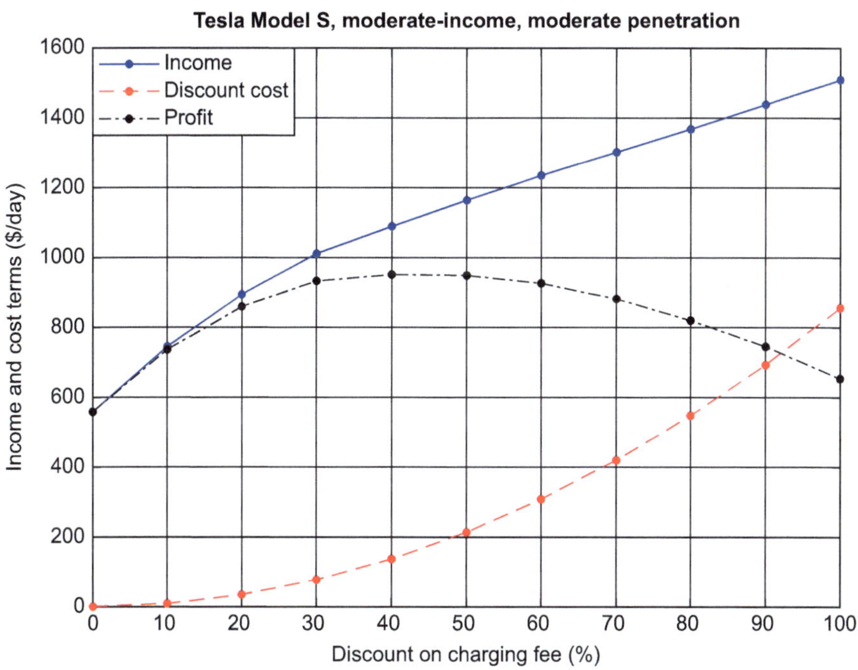

Fig. 5.30 The sensitivity analysis for the income, discount cost, and profit of parking lot with respect to the value of incentive considering Tesla Model S PEVs, moderate-income drivers, and moderate PEV penetration level

Table 5.6 The problem simulation results for Tesla Model S PEVs and high-income drivers considering different PEV penetration levels

	Before operation	After optimal operation of parking lot		
Outputs	Any penetration	Low penetration	Moderate penetration	High penetration
Optimal incentive (%)	0	100	90	90
Discount cost ($)	0	238	236	394
Income ($/day)	558	986	1017	1172
Maximum profit ($/day)	558	748	781	778
Enhancement ($/day)	–	189	222	219

problem simulation outcomes. As can be seen, the maximum profit of parking lot decreases for any PEV penetration level, compared to the previous cases. Despite the previous cases, herein, the moderate PEV penetration level leads to the most profit for the parking lot ($781/day). As can be noticed, the values of optimal incentive need to be updated and increased, and surprisingly the parking lot owner must charge the high-income drivers' vehicles in free (100% discount on charging fee) in the case of low PEV penetration level.

Figures 5.31, 5.32, and 5.33 illustrate the optimal hourly energy scheduling of energy sources of parking lot for each PEV penetration level. As can be seen, the energy sources of parking lot are rescheduled, due to the consideration of a different social class of drivers. In addition, as can be seen in Fig. 5.34, the sensitivity analysis of income, discount cost, and profit of parking lot with respect to the value of incentive demonstrates the existence of an optimal point for the value of incentive and parking lot's profit.

5.5.2.4 Sensitivity Analyses

Figures 5.35, 5.36, and 5.37 show the sensitivity analysis for the profit of parking lot with respect to the value of incentive for each social class of drivers considering Tesla Model S PEVs and various PEV penetration levels. As can be seen, the PEV penetration level and the social class of drivers are able to affect the profit and the maximum profit of parking lot. Therefore, the parking lot owner needs to study the social class of drivers and the PEV penetration level before solving the optimal operation problem of parking lot in the energy market. Otherwise, finding the optimal incentive and maximum profit of parking lot may be impossible, as can be realized from the figures.

The optimal value of incentive determined by the parking lot for each PEV penetration level and social class of drivers considering Tesla Model S PEVs is demonstrated in Fig. 5.38. As can be seen, the optimal value of incentive needs to be reinvestigated in each case. In addition, higher incentive is needed in the case of low PEV penetration level for any social class of drivers. Moreover, the high-income

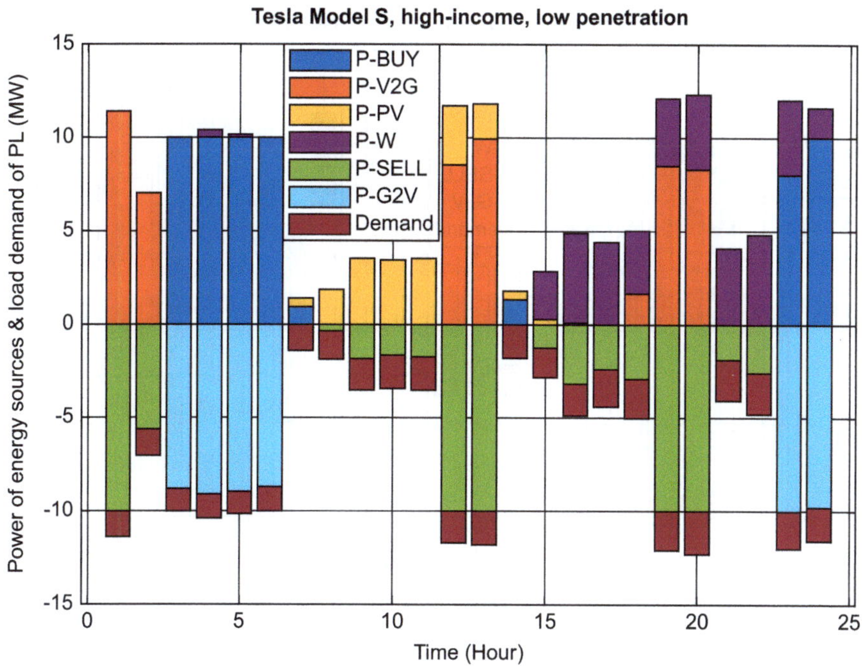

Fig. 5.31 The optimal hourly energy scheduling of different energy sources of parking lot considering Tesla Model S PEVs, high-income drivers, and low PEV penetration level

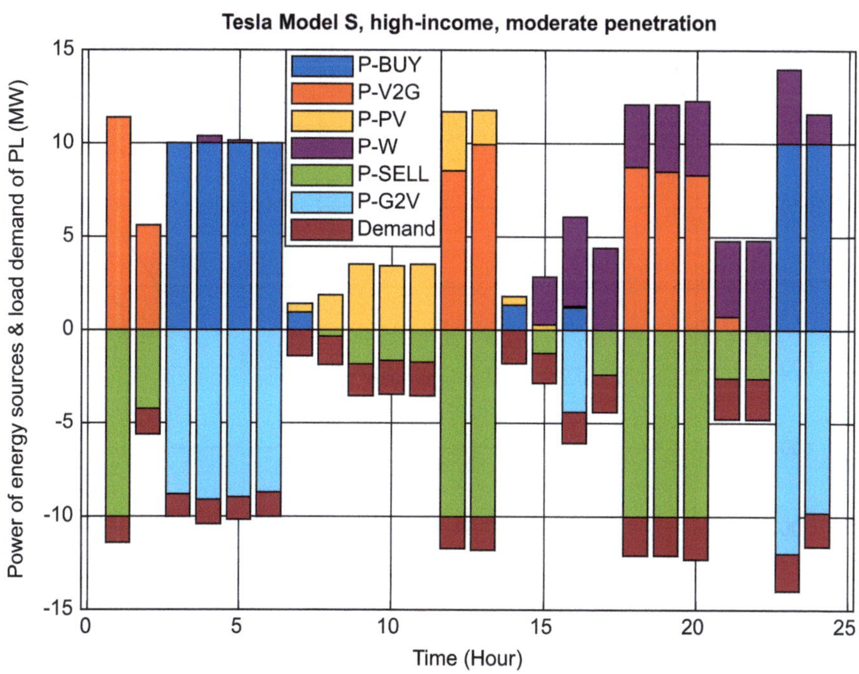

Fig. 5.32 The optimal hourly energy scheduling of different energy sources of parking lot considering Tesla Model S PEVs, high-income drivers, and moderate PEV penetration level

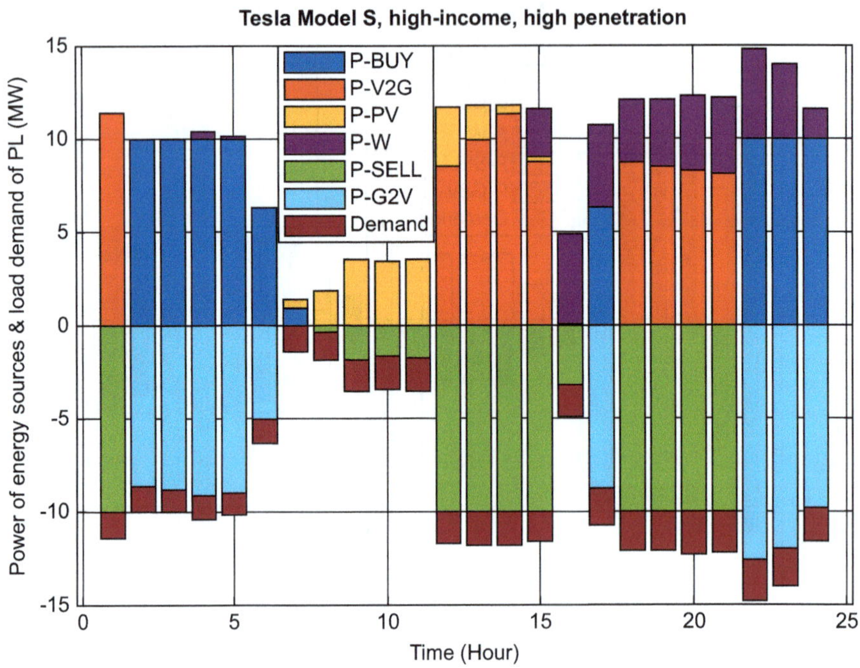

Fig. 5.33 The optimal hourly energy scheduling of different energy sources of parking lot considering Tesla Model S PEVs, high-income drivers, and high PEV penetration level

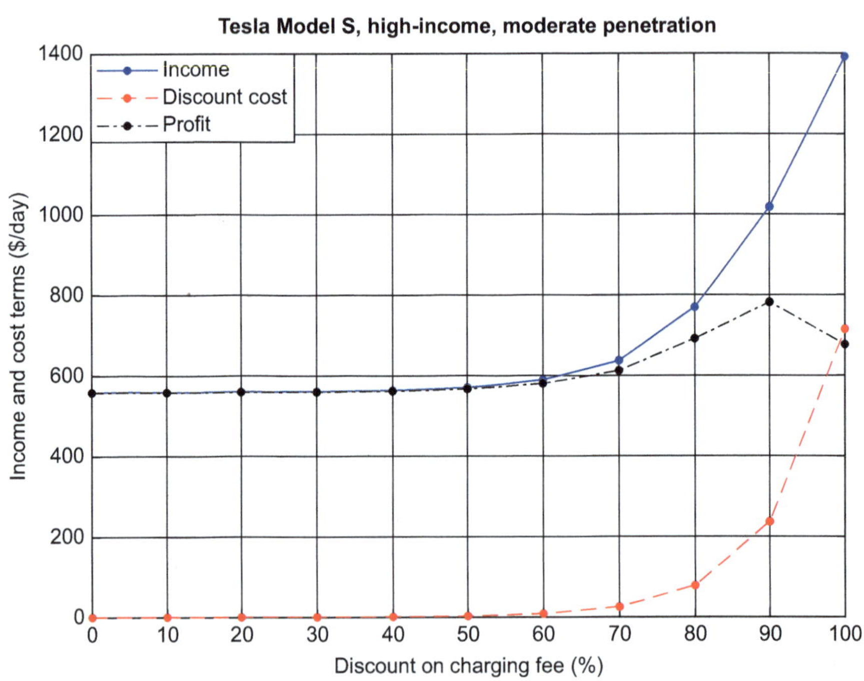

Fig. 5.34 The sensitivity analysis for the income, discount cost, and profit of parking lot with respect to the value of incentive considering Tesla Model S PEVs, high-income drivers, and moderate PEV penetration level

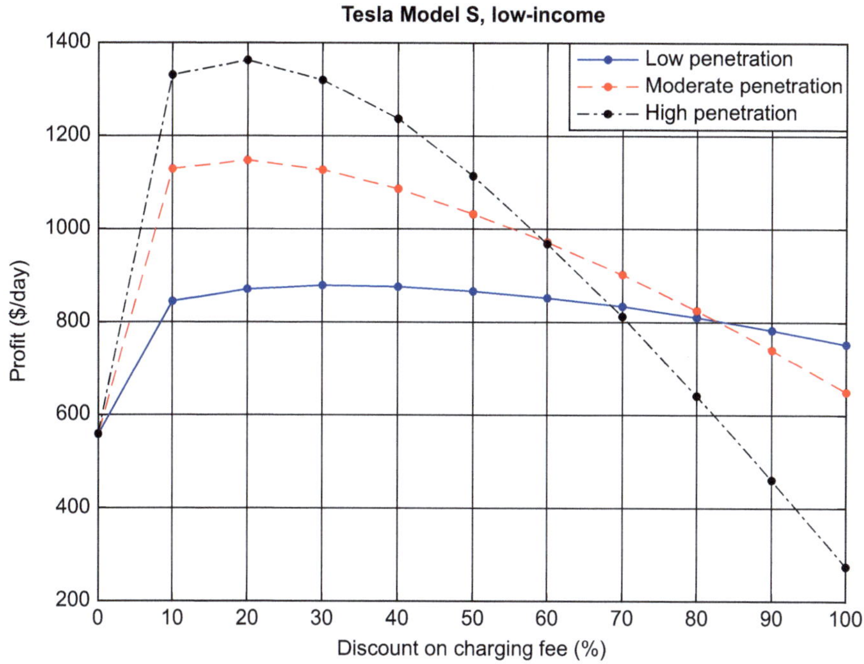

Fig. 5.35 The sensitivity analysis for the profit of parking lot with respect to the value of incentive for various PEV penetration levels considering Tesla Model S PEVs and low-income drivers

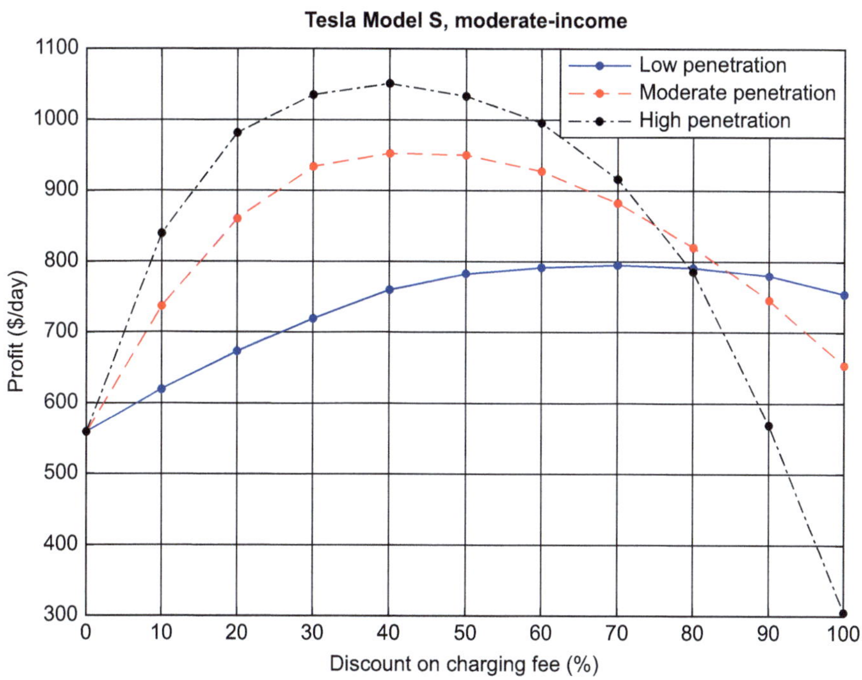

Fig. 5.36 The sensitivity analysis for the profit of parking lot with respect to the value of incentive for various PEV penetration levels considering Tesla Model S PEVs and moderate-income drivers

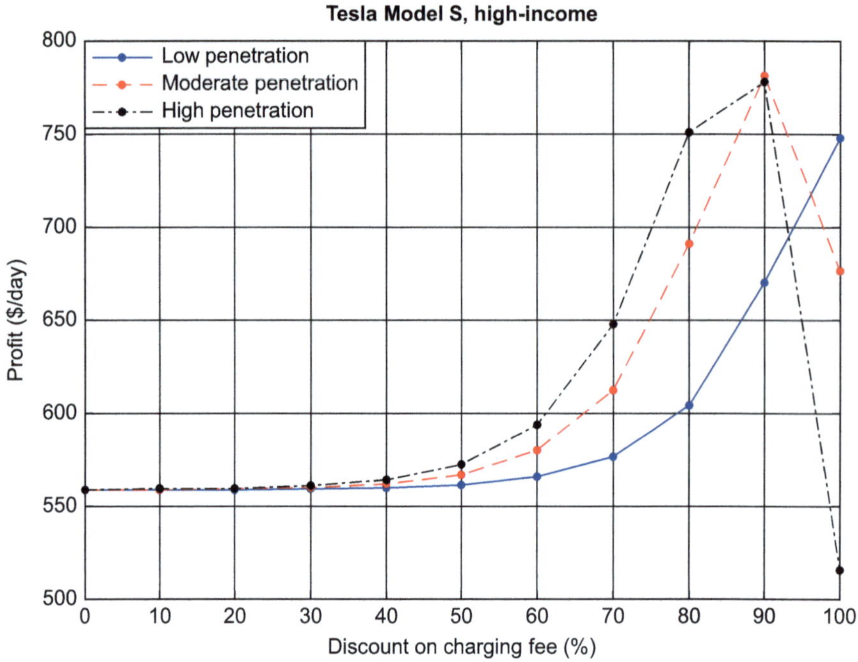

Fig. 5.37 The sensitivity analysis for the profit of parking lot with respect to the value of incentive for various PEV penetration levels considering Tesla Model S PEVs and high-income drivers

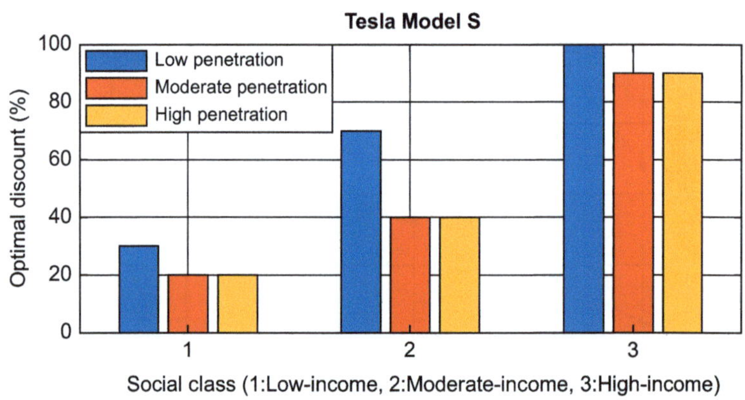

Fig. 5.38 The optimal value of discount on charging fee (incentive in percent) determined by the parking lot for each PEV penetration level and social class of drivers considering Tesla Model S PEVs

drivers need higher incentive in any PEV penetration level that implies the reluctance of high-income drivers with respect to the low amount of incentive.

Figure 5.39 displays the maximum profit of parking lot for each PEV penetration level and social class of drivers considering Tesla Model S PEVs. As can be seen, for

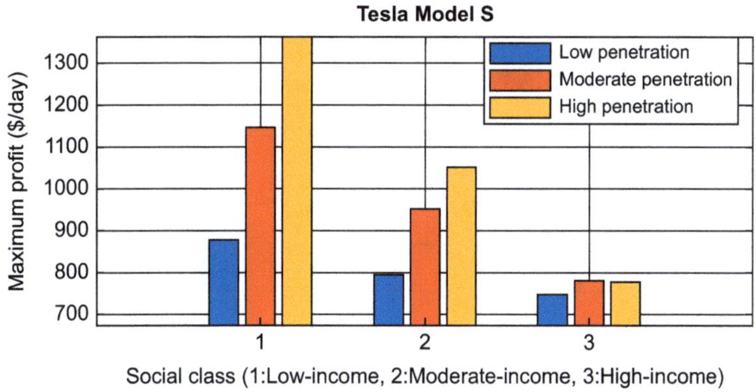

Fig. 5.39 The maximum profit of parking lot for each PEV penetration level and social class of drivers considering Tesla Model S PEVs

any PEV penetration level, the low-income and high-income social classes of drivers are the most and the least beneficial social classes for the parking lot owner, respectively. In addition, the case of high PEV penetration level results in the most profit of parking lot for the low-income and moderate-income social classes of drivers.

5.5.3 Studying the Other PEV Types

In this part of chapter, the maximum profit of parking lot, the optimal value of incentive, and the sensitivity analysis of parking lot's profit are studied for the other PEV types considering different social classes of drivers and PEV penetration levels.

Figure 5.40 shows the maximum profit of parking lot ($/day) for different PEV types, including Tesla Model S, Citroën C-Zero, Volkswagen e-Up, and Renault Kangoo Z.E., and for different PEV penetration levels and social classes of drivers. As can be seen, the type of PEV is able to affect the maximum profit of parking lot; thus it must be known in the problem of optimal operation of parking lot in the energy market. As can be seen, simulating the problem with Tesla Model S PEVs and Citroën C-Zero PEVs results in the most and the least maximum profits for the parking lot, respectively, for any social class of drivers and PEV penetration level.

The optimal values of incentive for different PEV types, PEV penetration levels, and social classes of drivers are illustrated in Fig. 5.41. As can be seen, the optimal value of incentive needs to be restudied in each case, since it might be affected by the abovementioned parameters. As can be noticed, the drivers of Tesla Model S and Citroën C-Zero need the least and the most incentives, respectively, for almost any social class of drivers and PEV penetration level.

Figures 5.42, 5.43, and 5.44 illustrate the sensitivity analysis for the parking lot's profit with respect to the value of incentive for different PEV types, PEV penetration levels, and social classes of drivers. As can be realized, to optimally operate the

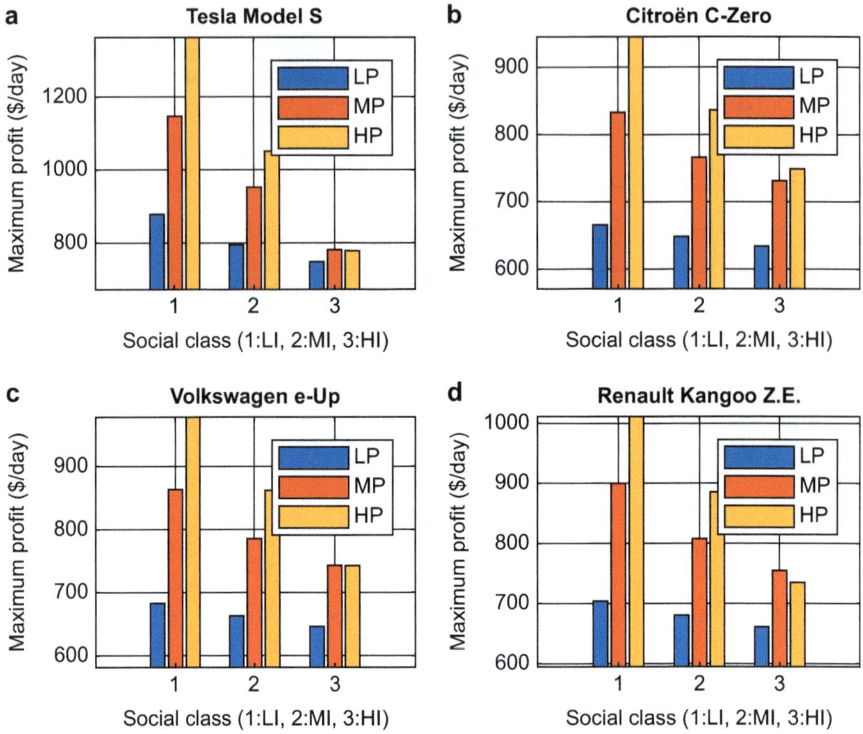

Fig. 5.40 The maximum profit of parking lot ($/day) for each PEV type, PEV penetration level (LP, low penetration; MP, moderate penetration; HP, high penetration), and social class of drivers (LI, low-income; MI, moderate-income; HI, high-income)

parking lot in the energy market, the type of PEVs, the social class of drivers, and the PEV penetration level need to be determined.

5.6 Conclusion

In this chapter, the drivers' behavior was modelled based on the social and geographical factors including the drivers' reaction with respect to the value of incentive (discount on charging fee), their social class (low-income, moderate-income, high-income), and their distance from the parking lot. The hourly driving routes of plug-in electric vehicles (PEVs) in the typical day were modelled considering the minimum and maximum traffic speed limits and the real latitude and longitude of area around

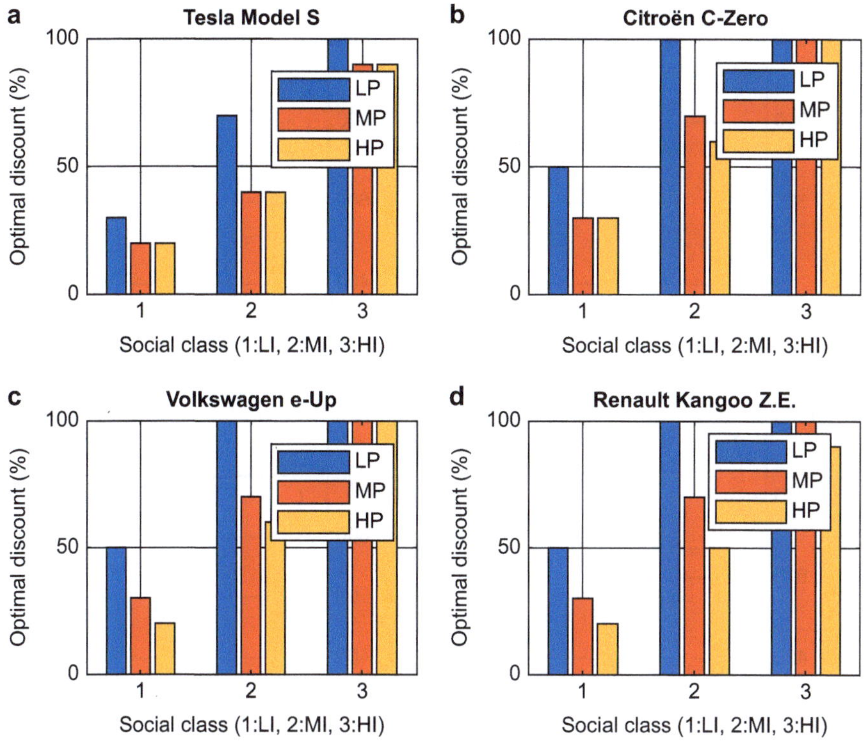

Fig. 5.41 The optimal value of discount on charging fee (incentive in percent) determined by the parking lot for each PEV type, PEV penetration level (LP, low penetration; MP, moderate penetration; HP, high penetration), and social class of drivers (LI, low-income; MI, moderate-income; HI, high-income)

Marina City vertical parking lot in Chicago, IL 60654. In addition, the arrival/departure time of PEVs to/from the parking lot and the state of charge (SOC) of PEVs upon their arrival to the parking lot were modelled. In this study, it was assumed that the parking lot is supplied by the renewable energy sources including wind turbines and photovoltaic panels.

The problem of optimal operation of parking lot in the energy market was formulated as a mixed integer linear programming (MILP) problem, and it was solved by MATLAB. The problem was solved for different PEV types including Tesla Model S, Citroën C-Zero, Volkswagen e-Up, and Renault Kangoo Z.E.; various PEV penetration levels such as low, moderate, and high penetration levels; as well as different social classes of drivers comprising low-income, moderate-income, and high-income.

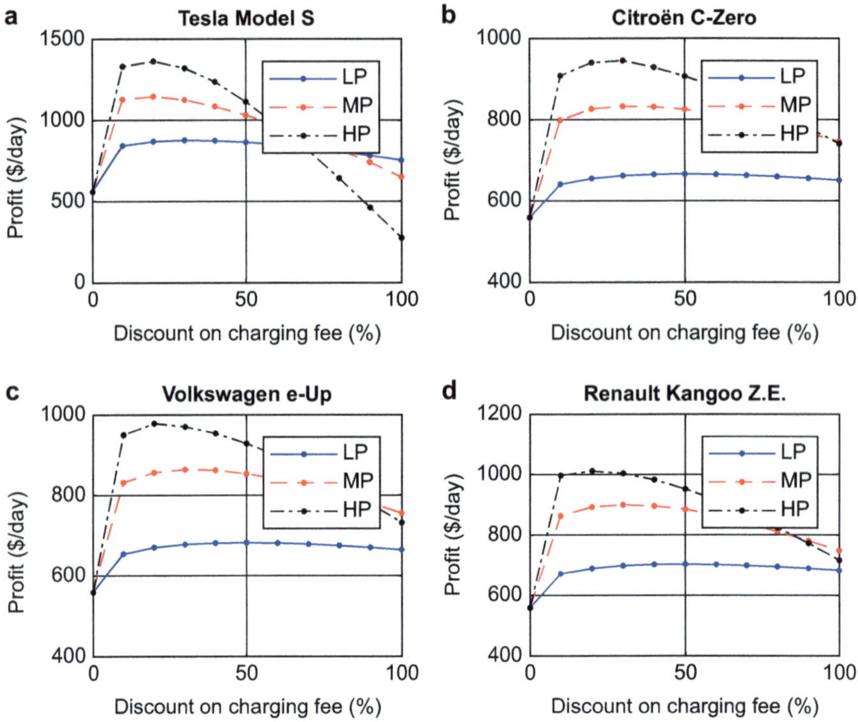

Fig. 5.42 The sensitivity analysis for the profit of parking lot with respect to the value of incentive for each PEV type and PEV penetration level (LP, low penetration; MP, moderate penetration; HP, high penetration) considering low-income drivers

It was seen that the social class of drivers, the type of PEV, and even the PEV penetration level are able to affect the problem simulation outcomes including the optimal operation of energy sources, the optimal value of incentive, and the maximum profit of parking lot. Thus, the parking lot owner must study and determine the type of abovementioned parameters before solving the optimal operation problem of parking lot in the energy market.

It was seen that the parking lot could make a profit in any case; however, the low-income and high-income drivers are the most and the least profitable drivers for the parking lot owner, respectively, for any PEV penetration level. It was observed that, to maximize the parking lot owner's profit, the most value of incentive needed to be considered for the high-income drivers and even the parking lot owner had to charge their vehicles in free in the case of low PEV penetration level.

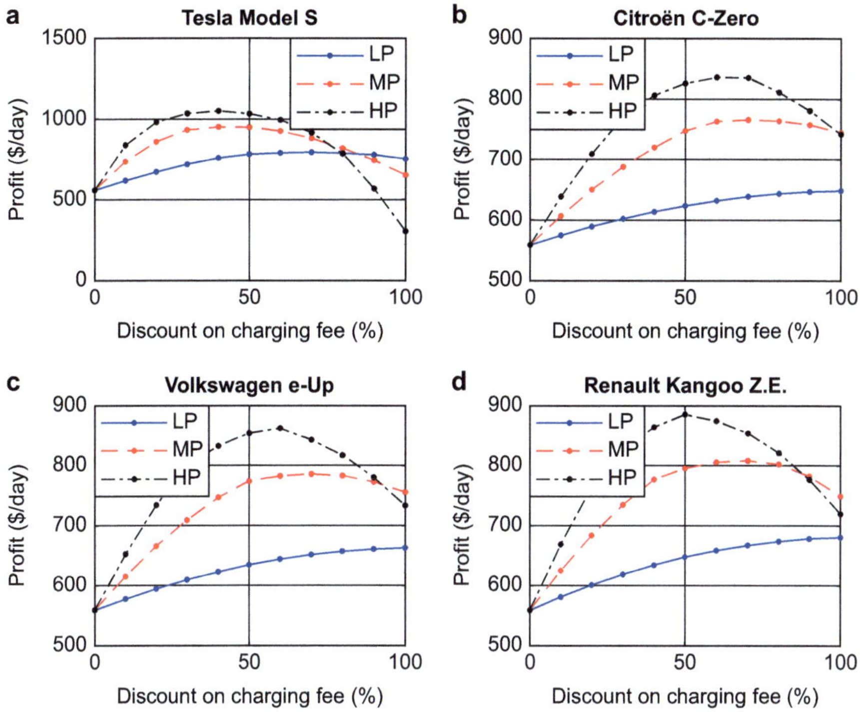

Fig. 5.43 The sensitivity analysis for the profit of parking lot with respect to the value of incentive for each PEV type and PEV penetration level (LP, low penetration; MP, moderate penetration; HP, high penetration) considering moderate-income drivers

In addition, it was noticed that the high PEV penetration level caused the most profit for the parking lot owner in most of the cases. Furthermore, it was proven that the drivers of Tesla Model S and Citroën C-Zero needed the least and the most incentives, respectively, for almost any social class of drivers and PEV penetration level. In addition, simulating the problem with these PEV types resulted in the most and the least maximum profits for the parking lot owner, respectively, for any social class of drivers and PEV penetration level.

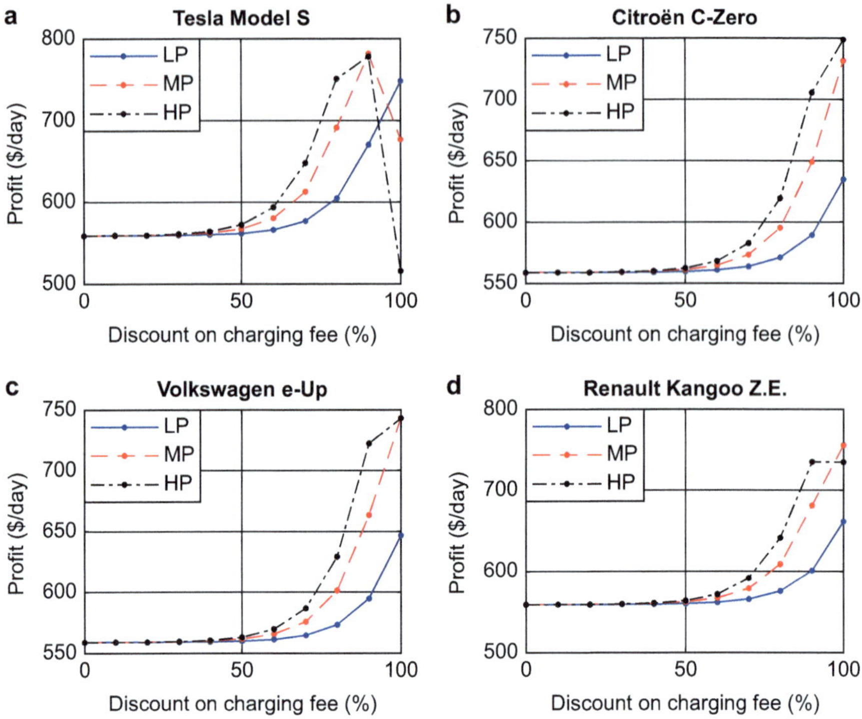

Fig. 5.44 The sensitivity analysis for the profit of parking lot with respect to the value of incentive for each PEV type and PEV penetration level (LP, low penetration; MP, moderate penetration; HP, high penetration) considering high-income drivers

Nomenclature

Problem parameters	
α	Distance of a PEV from the parking lot (meter)
β	Discount on charging fee of PEVs as the value of incentive (percent)
ξ	Responsiveness probability of drivers (percent)
π^{M}	Market electricity price ($/MWh)
π^{M-MIN}	Minimum price of electricity market ($/MWh)
ρ	Solar irradiance (W/m^2)
ρ^{st}	Solar irradiance in the standard environment (1000 W/m^2)
ρ^{c}	Certain solar irradiation point set as 150 W/m^2
η^{PV}	Efficiency of photovoltaic panels (percent)
A^{PV}	Total surface area of parking lot that covered by the photovoltaic panels (m^2)
Cap^{PEV}	Capacity of battery of a PEV (kWh)
C^{V2G}	Battery degradation cost of one PEV ($) due to the provision of V2G service for 1 hour

(continued)

C^{G2V}	Battery degradation cost of one PEV ($) due to the provision of G2V service for 1 hour
$Cost^{INC}$	Incentive cost ($)
$Cost^{BUY}$	Electricity purchasing cost ($)
$Cost^{V2G}$	Battery degradation cost of PEVs ($) due to the provision of V2G service
$Cost^{G2V}$	Battery degradation cost of PEVs ($) due to the provision of G2V service
DOD^{PEV}	Depth of discharge limit defined for a PEV (percent)
e	Index of PEV
Eff^{PEV}	Efficiency of a PEV's battery (percent) to provide G2V or V2G service
$Income^{SELL}$	Electricity purchasing income ($)
LI, MI, HI	Indices of low-income, moderate-income, and high-income social classes, respectively
n	Expected number of responsive drivers
N	Total number of PEVs in the area
OF	Objective function of problem ($)
P^{D}	Load demand of parking lot concerned with the lift stock installed in that
P^{BUY}	Purchasing power (MW)
P^{SELL}	Selling power (MW)
P^{V2G}	Power of PEVs during provision of V2G service (MW)
p^{V2G}	Power of a PEV during provision of V2G service (kW)
P^{G2V}	Power of PEVs during provision of G2V service (MW)
p^{G2V}	Power of a PEV during provision of G2V service (kW)
$\overline{P^{PEV}}$	Rated power of a PEV's battery (kW)
$\overline{P^{PL}}$	Nominal power of equipment installed in the parking lot as the maximum allowable power of parking lot (MW)
P^{PV}	Power of photovoltaic panels (MW)
$\overline{P^{PV}}$	Rated power of photovoltaic panels (MW)
P^{W}	Power of wind turbine (MW)
$\overline{P^{W}}$	Rated power of wind turbine (MW)
SC	Index of social class of drivers
SOC^{PEV}	State of charge of PEV (percent)
t, T	Index of time and the number of hours of operation period
v^{ci}	Cut-in wind speed (m/s)
v^{co}	Cut-out wind speed (m/s)
v	Wind speed (m/s)
v^{r}	Rated wind speed (m/s)
Y^{BUY}	Binary integer variable that indicates the purchasing status of parking lot
Y^{SELL}	Binary integer variable that indicates the selling status of parking lot
Y^{V2G}	Binary integer variable that indicates the status of PEV to provide V2G service
Y^{G2V}	Binary integer variable that indicates the status of PEV to provide G2V service
MILP matrices and vectors	

(continued)

$[f]$	Column vector of constants of objective function
$[X]$	Variables vector
$[A]$	Linear inequality constraints matrix
$[A_{eq}]$	Linear equality constraints matrix
$[b]$	Linear inequality constraints vector
$[b_{eq}]$	Linear equality constraints vector
$[l_b]$	Lower bounds vector
$[u_b]$	Upper bounds vector

Abbreviations

DOD	Depth of discharge
G2V	Grid-to-vehicle
LI, MI, HI	Low-income, moderate-income, high-income
LP, MP, HP	Low penetration, moderate penetration, high penetration
MILP	Mixed integer linear programming
PEV	Plug-in electric vehicle
SOC	State of charge
V2G	Vehicle-to-grid

References

1. W. Kempton et al., *A Test of Vehicle-to-Grid (V2G) for Energy Storage and Frequency Regulation in the PJM System* (University of Delaware, Newark, DE, 2008)
2. Canadian Automobile Association Electric Vehicles: What You Need to Know. http://electricvehicles.caa.ca/government-incentives/
3. According to BC Hydro's Draft Integrated Resource Plan, Appendix 2A–2011 Electric Load Forecast. https://www.bchydro.com/content/dam/hydro/medialib/internet/documents/environment/EVcharging_infrastructure_guidelines09.pdf
4. [Online]. Available: https://about.bnef.com/blog/lithium-ion-battery-costs-squeezed-margins-new-business-models/. Accessed on Jan 2019
5. [Online]. Available: https://about.bnef.com/blog/cumulative-global-ev-sales-hit-4-million/. Accessed on Jan 2019
6. [Online]. Available: https://about.bnef.com/blog/how-electric-cars-can-create-the-biggest-disruption-since-iphone/. Accessed on Jan 2019
7. W. Kempton, Vehicle to grid power. FERC (2007)
8. S. Han, S. Han, K. Sezaki, Development of an optimal vehicle-to-grid aggregator for frequency regulation. IEEE Trans. Smart Grid, 65–72 (2010)
9. M. Rahmani-andebili, Spinning reserve supply with presence of plug-in electric vehicles aggregator considering compromise between cost and reliability. IET Gener. Trans. Distrib. **7**, 1442–1452 (2013)
10. C. Jin, J. Tang, P. Ghosh, Optimizing electric vehicle charging with energy storage in the electricity market. IEEE Trans. Smart Grid **4**(1), 311–320 (2013)
11. F.J. Soares, P.M.R. Almeida, J.A.P. Lopes, Quasi-real-time management of electric vehicles charging. Elect. Power Syst. Res. **108**, 293–303 (2014)

12. M. Rahmani-Andebili, M. Fotuhi Firuzabad, M. Moeini-Aghtaie, Chapter 11: Optimal incentive plans for plug-in electric vehicles, in *Electric Distribution Network Planning*, (Springer Nature, Springer Singapore, 2018), pp. 299–320
13. M. Rahmani-Andebili, M. Fotuhi Firuzabad, An adaptive approach for PEVs charging management and reconfiguration of electrical distribution system penetrated by renewables. IEEE Trans. Ind. Inf. **14**(5), 2001 (2018)
14. M. Rahmani-Andebili, H. Shen, M. Fotuhi Firuzabad, Planning and operation of parking lots considering system, traffic, and drivers behavioral model. IEEE Trans. Syst. Man Cybern. Syst., 1 (2018). https://doi.org/10.1109/TSMC.2018.2824122
15. M. Rahmani-Andebili, Traffic and grid-based parking lot allocation for PEVs considering driver behavioral model, in *International Conference on Computing, Networking and Communications (ICNC)*, Silicon Valley, 26–29 Jan 2017
16. [Online]. Available: https://www.isp.state.il.us/traffic/speedlimitenf.cfm. Accessed on Jan 2019
17. [Online]. Available: https://www.popularmechanics.com/cars/g324/worlds-strangest-parking-garages/. Accessed on Jan 2019
18. [Online]. Available: http://www.elektriauto.ee/eng. Accessed on Jan 2019
19. [Online]. Available: https://www.buildinggreen.com/feature/folly-building-integrated-wind. Accessed on Jan 2019
20. [Online]. Available: http://www.solaripedia.com/13/117/cis_tower_solar_skyscraper_retrofit_(manchester,_uk).html. Accessed on Jan 2019
21. M. Rahmani-Andebili, Canopying plug-in electric vehicle parking lots with photovoltaic panels, in *IEEE Electric Power Distribution Networks Conference (EPDC)*, Karaj, pp. 1–4, 26–27 Apr 2016
22. [Online]. Available: http://www.wind-power-program.com. Accessed on Jan 2019
23. M. Rahmani-Andebili, G. K. Venayagamoorthy, Stochastic optimization for combined economic and emission dispatch with renewables, in *IEEE symposium series on computational intelligence*, Cape Town, pp. 1252–1258, 7 Dec 2015
24. MATLAB, R2014a. The Math Works, Inc.
25. R. Baldick, *Applied Optimization: Formulation and Algorithms for Engineering Systems* (Cambridge University Press, UK, 2009)
26. [Online]. Available: http://www.ferc.gov/market-oversight/mkt-electric/overview.asp. Accessed on Jan 2019

Chapter 6
Optimal Placement and Sizing of Parking Lots for the Plug-In Electric Vehicles Considering the Technical, Social, and Geographical Aspects

6.1 Introduction

Generally, there is a correlation between the incentives offered to the drivers by the governments and the adoption of electric vehicles and plug-in electric vehicles (PEV) [1]. The governments across the world are increasingly implementing the financial incentives at national, provincial, state, and even local level to speed up the transition from the internal combustion engine vehicles to the electric ones to achieve their own energy security and climate change mitigation goals.

Figure 6.1 shows the PEV share (percent) in some of the metropolitan cities of the USA with respect to the number of PEV promotion actions and the number of charging stations per million inhabitants [2]. As can be noticed, Atlanta, Portland, Seattle, and the four California state's major cities including San Francisco, San Diego, Los Angeles, and Riverside have the above-average adoption of PEV promotion actions and the PEV shares. In addition, most of the abovementioned cities have the above-average charging infrastructure. As can be seen in the graph, although Boston, Charlotte, Chicago, New York, Philadelphia, and Washington have the above-average adoption of PEV promotion actions, their PEV share is below the average value.

There are several charging standards for the PEVs including Level 1 Charging, Level 2 Charging, and DC Fast Charging. Level 1 Charging has been designed for home applications at voltage level of 120 V that has the lowest charging speed. Level 2 Charging standard (designed for both home and public applications) has the voltage level of 240 V that most of the common public chargers at offices, supermarkets, and parking lots use this charging standard [3]. The next one is DC Fast Charging that has the highest charging speed [3]. DC Fast Chargers include several noninterchangeable types, entitled, CHAdeMO, Combined Charging System (CCS), and Tesla's Supercharger. Most of the European and American automakers apply CCS, while CHAdeMO is utilized by the Japanese PEV manufacturers [3].

M. Rahmani-Andebili, *Planning and Operation of Plug-In Electric Vehicles*,
https://doi.org/10.1007/978-3-030-18022-5_6

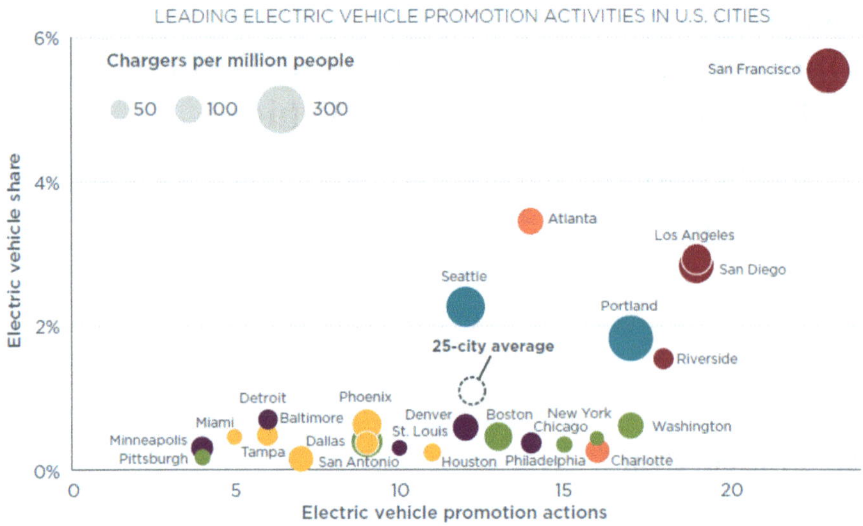

Fig. 6.1 The PEV share in some of the most populated US metropolitan cities with respect to the number of PEV promotion actions and the number of charging stations per million residents [2]

Figure 6.2a–e illustrates the location and distribution pattern of various types of charging stations in the USA [4]. In addition, Fig. 6.2f shows all the charging stations available in the USA. As can be seen in Fig. 6.2a, b, Level 1 and Level 2 charging stations are the least and the most available charging infrastructure in the USA, respectively. Moreover, Tesla DC Fast Charging stations are almost equally distributed across the country, as can be seen in Fig. 6.2e.

There are some papers in the literature that have studied the parking lot allocation and sizing problems [5–11]. In [6], the problem has been investigated in an electrical distribution system applying quantum-inspired simulated annealing optimization technique. In this reference, the parking lots are supplied by the photovoltaic panels, and they have the capabilities of generating and absorbing reactive power by changing the operating mode of their inverters. In other words, the solar parking lots and the PEVs are utilized as the auxiliary sources of active and reactive powers by the local distribution company (DISCO). In [7], the parking lots are considered as the energy storage during the self-healing process. In [8], the parking lots have been placed in the system to improve its reliability. However, in [5–8], the geographical and social aspects of problem have been ignored.

In [9], the problem has been defined to improve the system reliability and facilitate the participation of DISCO in the energy market transactions. In this reference, just the behavior of drivers with respect to the value of incentive has been modelled. In [10, 11], the parking lot allocation problem has been solved considering the driving pattern of PEVs' owners and modelling the drivers' reaction with respect to the value of incentive and their distance from the parking lot. However, in these studies, the dataset of driving routes is not real, and the feeder's failure rate (FFR) and voltage-dependent load (VDL) have not been modelled in the

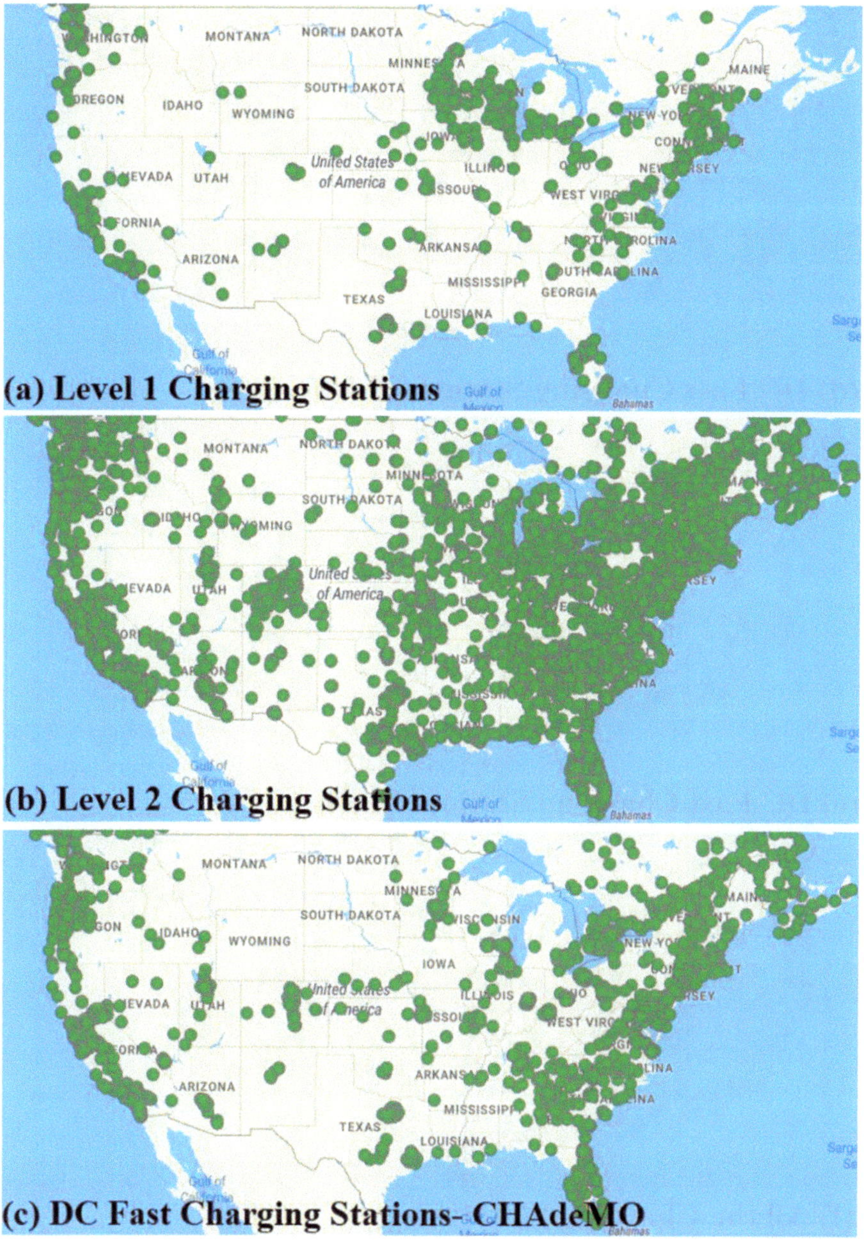

Fig. 6.2 The location and distribution pattern of various types of charging stations in the USA [4]

problem. Moreover, the sensitivity analysis regarding the effects of different parameters on the problem outputs has not been performed.

In this chapter, the planning problem of parking lot sizing and allocation is investigated in an electrical distribution system considering the security constraints of grid such as voltage magnitude bounds of buses and power magnitude limit of

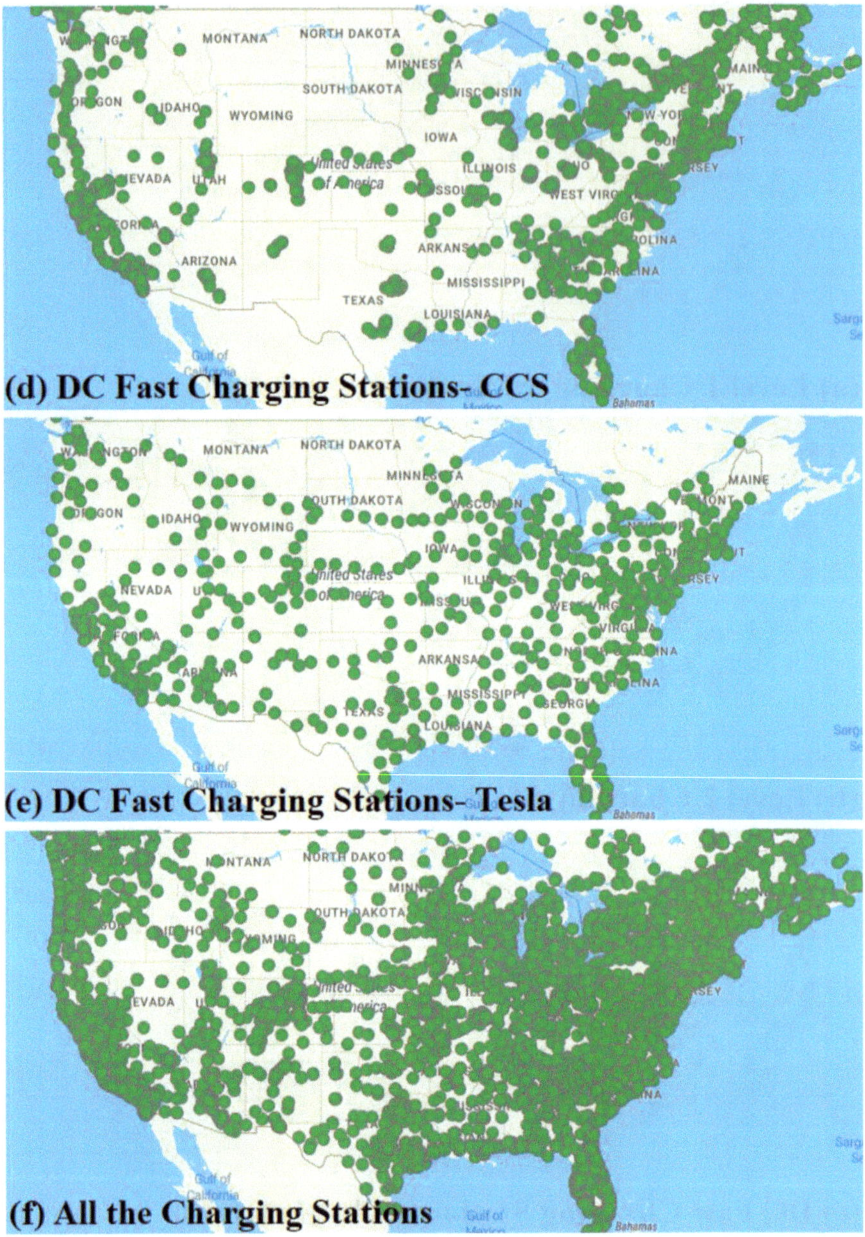

Fig. 6.2 (continued)

branches, as well as modelling the technical, geographical, and social aspects of problem. Herein, the goal of local DISCO is to minimize the total cost of planning problem during the given planning period. The planning problem has several cost terms that include:

- The investment cost to install the parking lots in the electrical distribution network and equip them with the charging infrastructure
- The present worth value of maintenance cost of parking lots and their equipment during the planning period
- The present worth value of incentives paid to the drivers to motivate them to provide the vehicle-to-grid (V2G) service in the proposed locations of system during the planning period
- The present worth value of energy loss cost of feeder's branches during the planning time horizon
- The present worth value of ENS cost due to the faults occurred in the feeder's branches during the planning period

Herein, in order to calculate the present worth value of cost terms, the economic factors that include the inflation and interest rates are applied. In addition, in this study, the real driving routes of vehicles in San Francisco are considered. Moreover, the FFR and VDL (related to the electricity consumer types including residential, commercial, and industrial) are modelled in the study.

In this chapter, in addition to the investigation of primary and final statuses of system and problem, several scenarios are defined to study the effects of different social classes of drivers (low-income, moderate-income, and high-income), the PEV penetration levels (low, moderate, and high), the PEV types (Citroën C-Zero and Tesla Model S), the FFR models (constant, linear, power, exponential, and logarithmic), and the VDL models (residential, commercial, and industrial) on the planning problem outputs. Herein, the problem outputs include the optimal hourly location and size of parking lots, the optimal hourly value of incentive, the minimum total cost of planning problem, the energy loss and ENS of system during the planning period, the reliability indices of system, the voltage profile of buses, and the apparent power flowing through the branches of feeder.

In the following, the geographical and social aspects of PEVs and drivers are modelled in Sect. 6.2. The VDL and FFR are modeled in Sects. 6.3 and 6.4, respectively. In Sects. 6.5 and 6.6, the problem formulation and the optimization technique are presented, respectively. Section 6.7 is concerned with the problem simulation, and the conclusion is presented in Sect. 6.8.

6.2 Modelling the Geographical and Social Aspects of Problem

6.2.1 Modelling Drivers' Behavior

The drivers, who are responsive to the local DISCO to provide the V2G service at the suggested parking lots and proposed time, will be paid an incentive, in addition to the V2G service provision cost. In this chapter, a certain percentage of V2G service provision cost (β) is defined as the value of incentive. Herein, the responsiveness probability of a driver is modelled based on his/her social class (low-income, moderate-income, high-income) and his/her distance from the parking lot [10–12]. Moreover, it is rationally supposed that the low-income and high-income drivers are the most and the least responsive drivers, since they are more and less likely to be interested in the proposed incentive. Therefore, a power function and an exponential function are applied to model their behavior with respect to the value of incentive and their distance from the parking lot. Moreover, a linear function is used to model the behavior of moderate-income drivers.

Table 6.1 presents the responsiveness probability function of a low-income, moderate-income, and high-income driver with respect to the value of incentive (β in percent) and his/her distance from the parking lot (α in km) for the ranges of $0 \leq \beta \leq 100$ percent and $0 \leq \alpha \leq 10$ km. Herein, *LI*, *MI*, and *HI* are the indices of low-income, moderate-income, and high-income drivers, respectively. There are some logics in these models presented as follows:

- If no incentive is offered to the driver ($\beta = 0\%$), the driver, with any social class and any distance from the parking lot, will not go toward the parking lot to provide the V2G service.
- The driver, with any social class, will head for the parking lot to provide the V2G service if he/she has the nearest distance from the parking lot ($\alpha \leq 1$ km) and the value of incentive is 100%.
- The driver, with any social class, will not be motivated by any incentive to provide the V2G service if the parking lot is too far ($\alpha > 10$ km).
- The driver will respond to other cases for the ranges of $0 \leq \beta \leq 100$ percent and $0 \leq \alpha \leq 10$ km with the probability value presented in Table 6.1 and graphically illustrated in Figs. 6.3, 6.4, and 6.5.

As can be seen in Figs. 6.3, 6.4, and 6.5, the whole three-dimensional space is logically covered by the surfaces; thus the models approximately include all the possibilities of a driver's reaction. As can be seen in Fig. 6.3, a low-income driver has a relatively remarkable responsiveness with respect to a small value of incentive (e.g., 10%) even from the farthest distance (e.g., 9 km). However, as can be seen in Fig. 6.5, a high-income driver does not have any reaction with respect to a relatively large value of incentive (e.g., 50%) even from the closest distance (1 km).

To convert the distance travelled by a PEV (in degree) to a quantity in km in San Francisco, CA, USA, the latitude and longitude values (degree) need to be multiplied by 111.03 and 85.39, respectively [13].

Table 6.1 The responsiveness probability of a low-income (LI), moderate-income (MI), and high-income (HI) driver with respect to the value of incentive in percent (β) and his/her distance from the parking lot in km (α)

Social class	Driver's responsiveness probability (percent)
Low-income (LI)	$\xi_{\alpha,\beta}^{LI} = 100 \times \left(1 - \frac{\alpha}{10}\right)^{0.3} \times \left(\frac{\beta}{100}\right)^{0.3}$ (6.1)
Moderate-income (MI)	$\xi_{\alpha,\beta}^{MI} = 100 \times \left(1 - \frac{\alpha}{10}\right) \times \left(\frac{\beta}{100}\right)$ (6.2)
High-income (HI)	$\xi_{\alpha,\beta}^{HI} = 100 \times exp\left(-10 \times \frac{\alpha}{10}\right) \times exp\left(10 \times \left(\frac{\beta}{100} - 1\right)\right)$ (6.3)

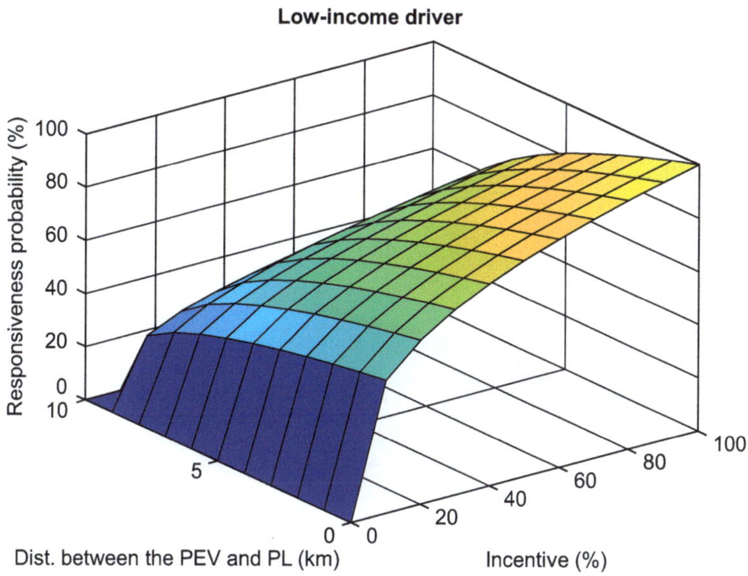

Fig. 6.3 The responsiveness probability of a low-income driver with respect to the value of incentive and his/her distance from the parking lot (PL) for the ranges of $0 \leq \beta \leq 100$ percent and $0 \leq \alpha \leq 10$ km

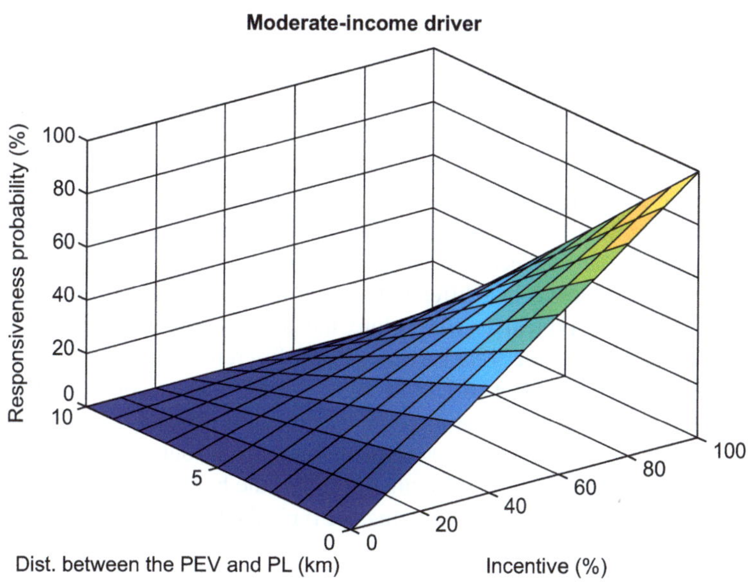

Fig. 6.4 The responsiveness probability of a moderate-income driver with respect to the value of incentive and his/her distance from the parking lot (PL) for the ranges of $0 \leq \beta \leq 100$ percent and $0 \leq \alpha \leq 10$ km

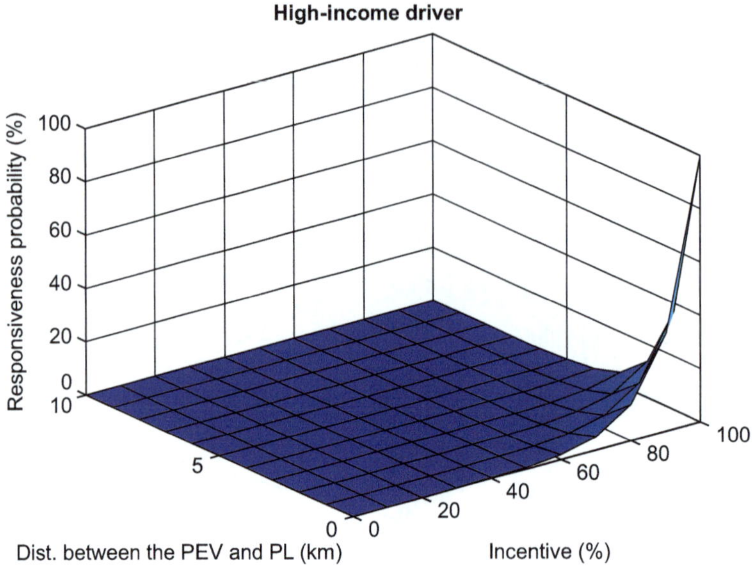

Fig. 6.5 The responsiveness probability of a high-income driver with respect to the value of incentive and his/her distance from the parking lot (PL) for the ranges of $0 \leq \beta \leq 100$ percent and $0 \leq \alpha \leq 10$ km

6.2.2 Driving Routes in San Francisco

In this study, the actual driving routes of vehicles in San Francisco, CA, USA, are considered [14]. Herein, it is assumed that the PEVs have the same routes of conventional vehicles. Figures 6.6, 6.7, and 6.8 illustrate eight hourly driving routes in San Francisco during the typical day [14]. These figures show the hourly real latitude and longitude of location of PEVs, in addition to the real latitude and longitude of location of electrical distribution network. As can be seen, in each driving route, the vehicle is in a different location; thus its distance from any parking lot changes during the typical day. Therefore, the driver's responsiveness probability to the DISCO to provide the V2G service will be different during the day, based on the models presented in Sect. 6.2.1. In other words, even a constant amount of incentive will result in a dissimilar driver's reaction during the day. This implies that the local DISCO needs to model or estimate the driving routes, in addition to the modelling of drivers' social class. Figure 6.9 shows the hourly geographical driving routes in San Francisco, CA, USA, during the typical day.

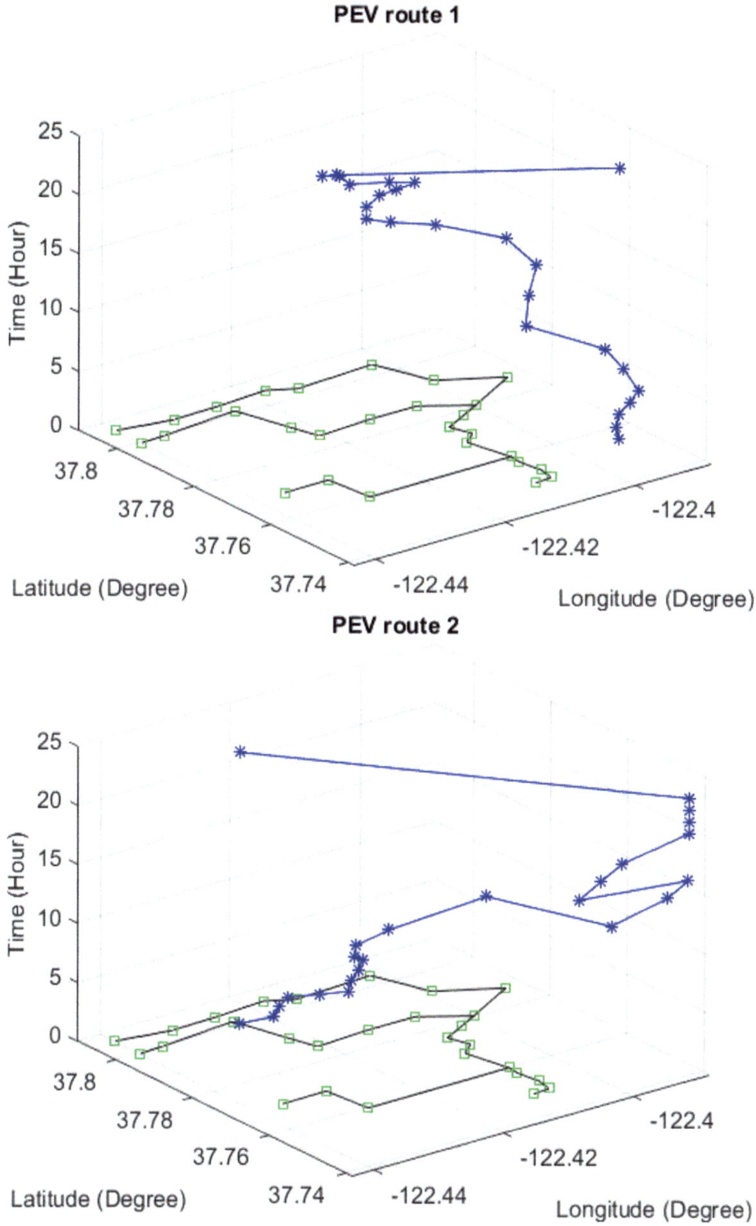

Fig. 6.6 The hourly driving routes 1–3 in San Francisco, CA, USA, during the typical day [14] and the real latitude and longitude of location of electrical distribution network

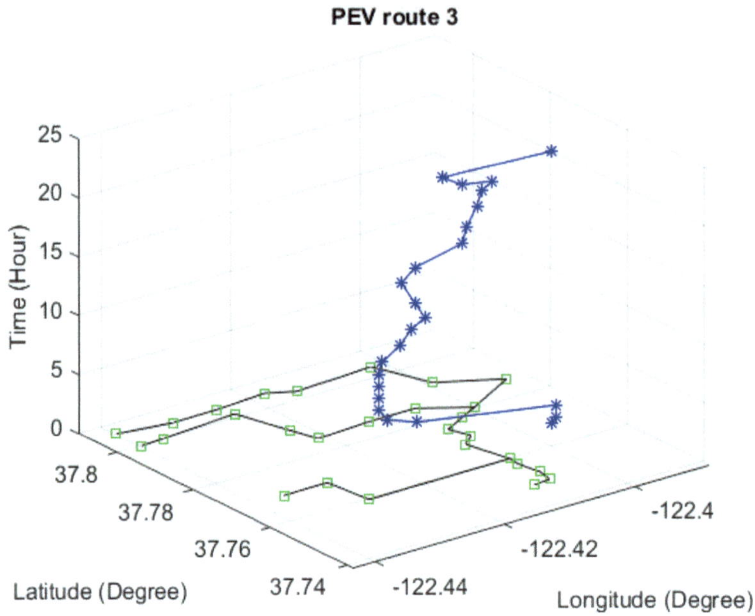

Fig. 6.6 (continued)

6.3 Modelling Voltage-Dependent Load

In a real electrical distribution system, the voltage profile of a bus is not equal to its rated value, due to the power and energy that are lost in the electrical feeder. Therefore, the active and reactive power demands of an electricity consumer will change, since they depend on the operating voltage profile of bus [15–19]. In this study, the VDL is modelled in the PQ-buses, and the value of active and reactive power demands are assumed as the dependent variables in the power flow problem, in addition to the independent variables including the voltage magnitude and voltage phase angle.

The relations (6.4) and (6.5) present the power flow equations of PQ-buses, where P and Q are the unknown dependent variables that need to be updated using the VDL model presented in (6.6) and (6.7). Herein, P, Q, V, Y, φ, δ, i, j, J, and N^J are the operating active power demand, the operating reactive power demand, the operating voltage profile of bus, the admittance of line between two buses, the value of phase angle of admittance in polar coordinate system, the value of voltage's phase angle of a bus in polar coordinate system, the index of bus, the index of bus, the set of buses, and total number of buses of system, respectively. In addition, P_0, Q_0, and V_0 are the rated active power demand, the rated reactive power demand, and the rated voltage profile of bus, respectively.

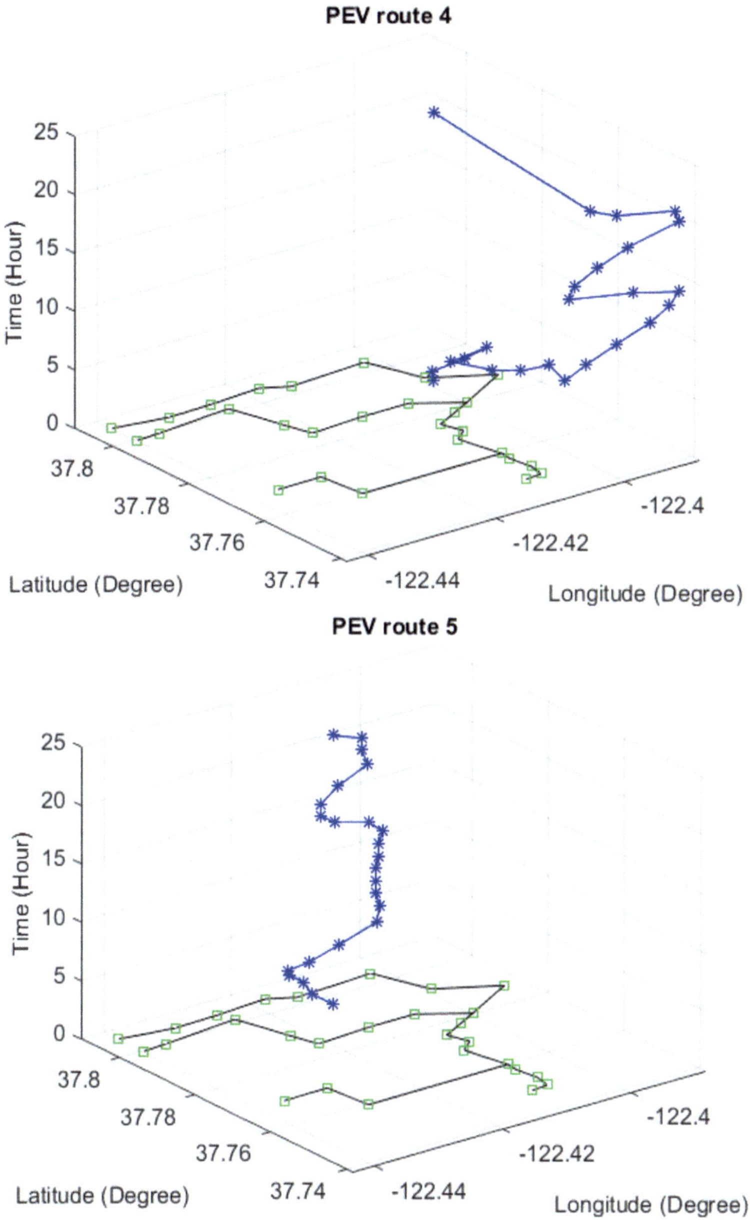

Fig. 6.7 The hourly driving routes 4–6 in San Francisco, CA, USA, during the typical day [14] and the real latitude and longitude of location of electrical distribution network

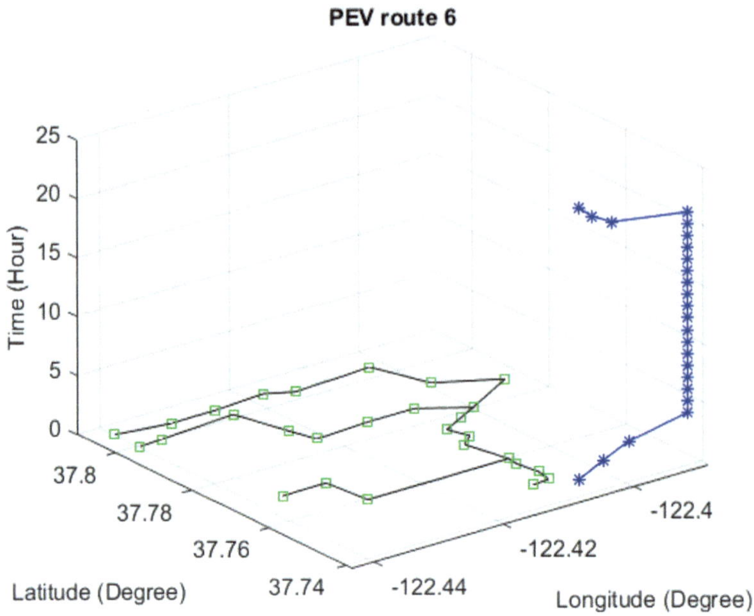

Fig. 6.7 (continued)

The parameters ρ_1 and ρ_2 are the exponents of active and reactive power demands of an electricity consumer in the VDL model, respectively [19]. Table 6.2 presents these exponents for different types of electricity consumer [19].

$$P_i = \sum_{j \in J} |V_i||V_j||Y_{ij}| \cos\left(\varphi_{ij} - \delta_i + \delta_j\right), \quad \forall i \in J, J = \left\{1, \ldots, N^J\right\} \tag{6.4}$$

$$Q_i = -\sum_{j \in J} |V_i||V_j||Y_{ij}| \sin\left(\varphi_{ij} - \delta_i + \delta_j\right), \quad \forall i \in J \tag{6.5}$$

$$P_i = P_{0,i}\left(\frac{|V_i|}{|V_{0,i}|}\right)^{\rho_1}, \quad \forall i \in J \tag{6.6}$$

$$Q_i = Q_{0,i}\left(\frac{|V_i|}{|V_{0,i}|}\right)^{\rho_2}, \quad \forall i \in J \tag{6.7}$$

6.4 Modelling Feeder's Failure Rate

Distributed generation in an electrical distribution system can locally supply part of the active power demand of consumers and decrease the magnitude of apparent power flowing through the feeder. Therefore, the magnitude of current flowing

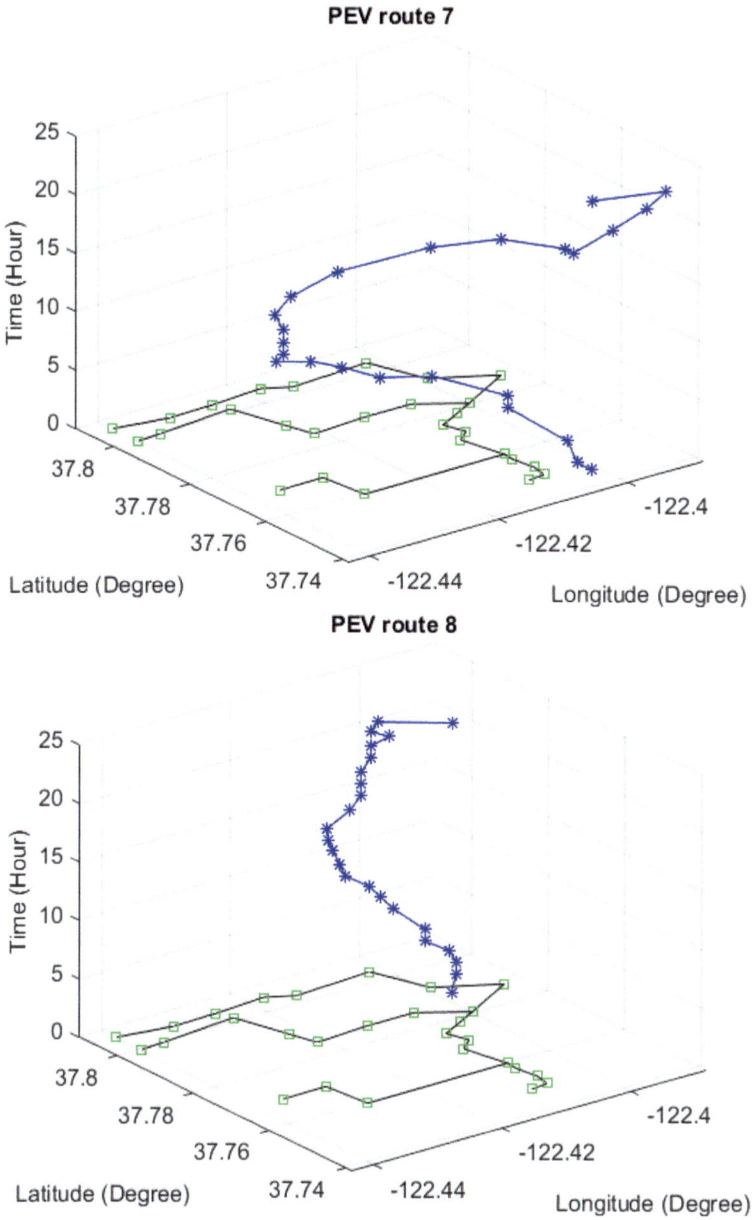

Fig. 6.8 The hourly driving routes 7–8 in San Francisco, CA, USA, during the typical day [14] and the real latitude and longitude of location of electrical distribution network

Fig. 6.9 The hourly geographical driving routes in San Francisco, CA, USA, during the typical day [14]

through the feeder will reduce, the active and reactive power losses will decrease, and consequently the temperature of feeder will drop [15–17].

Table 6.2 The exponent of active and reactive power demands concerned with different electricity consumer types [19]

Electricity consumer type	ρ_1	ρ_2
Residential (Res)	0.92	4.04
Commercial (Com)	1.51	3.40
Industrial (Ind)	0.18	6.00

Table 6.3 FFR models as the function of variable $Real\{I\}$ and constant parameters including $Real\{I_0\}$, λ_0, and λ_f [15–17]

FFR model	Mathematical expression of FFR models
Constant (CON)	$\lambda^{CON} = \lambda_0$ (6.8)
Linear (LIN)	$\lambda^{LIN} = (\lambda_0 - \lambda_f) \times \left(\frac{Real\{I\}}{Real\{I_0\}}\right) + \lambda_f$ (6.9)
Power (POW)	$\lambda^{POW} = \left(((\lambda_0)^n - (\lambda_f)^n) \times \left(\frac{Real\{I\}}{Real\{I_0\}}\right) + (\lambda_f)^n\right)^{\frac{1}{n}}, n = 0.5$ (6.10)
Exponential (EXP)	$\lambda^{EXP} = exp\left(\left(\ln(\lambda_0) - \ln(\lambda_f)\right) \times \left(\frac{Real\{I\}}{Real\{I_0\}}\right) + \ln(\lambda_f)\right)$ (6.11)
Logarithmic (LOG)	$\lambda^{LOG} = \ln\left(\left(e^{\lambda_0} - e^{\lambda_f}\right) \times \left(\frac{Real\{I\}}{Real\{I_0\}}\right) + e^{\lambda_f}\right)$ (6.12)

Therefore, optimal V2G service provision, as the specific type of distributed generation, in the electrical distribution system can decrease the FFR, since the high temperature of feeder has a destructive effect on itself such as causing the overhead lines to sag and insulation problems in the underground cables [20, 21]. In other words, it is necessary to study the positive effects of optimal V2G service provision on the electrical distribution system, caused by the FFR reduction. Some of these positive effects include the system power loss reduction and reliability improvement.

In this study, it is assumed that a branch of feeder, initially carrying the current with the real component of $Real\{I_0\}$, has the primary FFR of λ_0. However, after fully compensating the active power demand of two buses concerned with the branch, the real component of current flowing through the branch is removed ($Real\{I\} = 0$), the magnitude of current is minimized, and consequently the failure rate of branch decreases to λ_f (minimum failure rate).

Herein, four mathematical relations including linear (LIN), power (POW), exponential (EXP), and logarithmic (LOG) functions are applied to formulate the FFR models. Table 6.3 presents these FFR models that are the functions of $Real\{I\}$, as the variable, and $Real\{I_0\}$, λ_0, and λ_f, as the constant parameters. In addition, the constant model of FFR (CON) is also presented in Table 6.3 that ignores the dependency of FFR on the real component of current flowing through the branch. As can be seen, in each model, a branch of feeder will have a failure rate of λ_0 if it carries the real component of current of $Real\{I_0\}$, and its failure rate will change to λ_f if its real component of current is zero.

In this chapter, the value of λ_f is considered about $0.1 \times \lambda_0$ for each branch of feeder. Figure 6.10 shows the relative value of failure rate of a branch ($\lambda^{Model}/\lambda_0$)

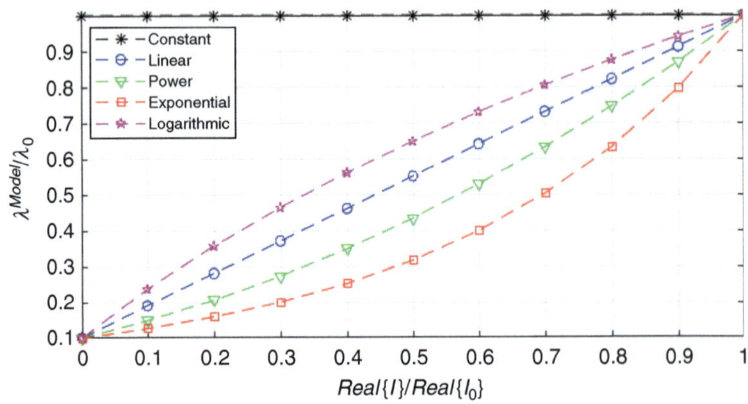

Fig. 6.10 Relative value of failure rate of a branch with respect to the relative value of real component of current flowing through the branch for different models ($\lambda^{Model} \epsilon \{\lambda^{CON}, \lambda^{LIN}, \lambda^{POW}, \lambda^{EXP}, \lambda^{LOG}\}$)

with respect to the relative value of real component of current flowing through the branch ($Real\{I\}/Real\{I_0\}$) for different models. As can be seen, in all the models, if the real component of current flowing through the branch ($Real\{I\}$) is zero, the value of branch's failure rate (λ^{Model}) is equal to $0.1 \times \lambda_0$ (or λ_f). Moreover, the value of branch's failure rate (λ^{Model}) is equal to its initial value (λ_0) if the real component of current flowing through the branch ($Real\{I\}$) is equal to its primary value ($Real\{I_0\}$). However, as is seen, for other changes of real component of current, the branch's failure rate is different for each model. Herein, the constant model, as the conventional model, does not have any sensitivity with respect to the changes of real component of current flowing through the branch.

Due to the dependency of FFR on the real component of current flowing through the branch, as well as, the model, the reliability indices need to be revised. As can be seen in Eqs. (6.13, 6.14, 6.15, 6.16, and 6.17), the reliability indices are modified due to the updated value of λ_i^{Model} and U_i^{Model}.

Equation (6.13) presents the system average interruption frequency index (SAIFI) in failure/year (f/y) per customer. Herein, N^C is the number of customers connected to the load point. The system average interruption duration index (SAIDI) in hour/year (h/y) per customer is presented in (6.14), where U^{Model} is the annual outage time of a load point. Equation (6.15) presents the energy not supplied (ENS) index in MWh/year, where LNS is the load not supplied of load point during the fault occurrence interval. The average energy not supplied (AENS) index in MWh/year per customer is presented in (6.16). In addition, the average system availability index (ASAI) in percent is given in (6.17).

$$SAIFI = \frac{\sum_{i \in J} \lambda_i^{Model} \times N_i^C}{\sum_{i \in J} N_i^C}, \lambda^{Model} \epsilon \{\lambda^{CON}, \lambda^{LIN}, \lambda^{POW}, \lambda^{EXP}, \lambda^{LOG}\} \tag{6.13}$$

$$SAIDI = \frac{\sum_{i \in J} U_i^{Model} \times N_i^C}{\sum_{i \in J} N_i^C}, \lambda^{Model} \epsilon \{\lambda^{CON}, \lambda^{LIN}, \lambda^{POW}, \lambda^{EXP}, \lambda^{LOG}\} \tag{6.14}$$

$$ENS = \sum_{i \in J} U_i^{Model} \times LNS_i, \lambda^{Model} \epsilon \{\lambda^{CON}, \lambda^{LIN}, \lambda^{POW}, \lambda^{EXP}, \lambda^{LOG}\} \tag{6.15}$$

$$AENS = \frac{ENS}{\sum_{i \in J} N_i^C} \tag{6.16}$$

$$ASAI = \left(1 - \frac{SAIDI}{8760}\right) \times 100 \tag{6.17}$$

6.5 Problem Formulation

6.5.1 Objective Function

The objective function of problem is minimizing total cost of planning problem during the planning time horizon. In this study, the economic factors such as inflation and interest rates are applied to consider the present worth value of cost terms. As can be seen in (6.18), the objective function of problem ($) comprises five cost terms including investment cost to build the parking lots and equip them with the charging/discharging stations ($Cost^{INV}$), the present worth value of maintenance cost of parking lots and their equipment during the planning period ($PWV(Cost^{M})$), the present worth value of incentives paid to the drivers to motivate them to provide the V2G service during the planning interval ($PWV(Cost^{INC})$), the present worth value of energy loss cost of feeder's branches during the planning time horizon ($PWV(Cost^{LOSS})$), and the present worth value of ENS cost due to the faults and failures occurred in the feeder's branches during the planning period ($PWV(Cost^{ENS})$).

$$OF = min\{Cost^{INV} + PWV(Cost^{M}) + PWV(Cost^{INC}) + PWV(Cost^{LOSS}) + PWV(Cost^{ENS})\} \tag{6.18}$$

Herein, the load of electricity consumers including residential, commercial, and industrial customers is modelled as the VDL in the Newton-Raphson power flow problem. Moreover, an hourly varying pattern is considered for the system load. In addition, different hourly electricity prices and ENS costs are assumed for various types of customers. Furthermore, to evaluate the system reliability and calculate the

ENS cost, the FFR is modelled using some mathematical functions including linear, exponential, power, and logarithmic models.

6.5.2 *Cost Terms*

6.5.2.1 Investment Cost

The total investment cost ($) to build the parking lots and equip them with the charging/discharging stations is presented in (6.19). As can be seen, it depends on the building cost of a parking lot and equipping it just for one PEV (C^{PL} in $), the size (capacity) of parking lots ($Size^{PL}$) installed in different buses. Equation (6.20) indicates that the size of a parking lot must change with a certain step. Herein, b, B, N^B, n^{PL}, $Size^{PL-MIN}$, and n^{PL-MAX} are the index of branch, the set of branches of feeder, the total number of branches, the step to change the size of a parking lot, the minimum size of a parking lot, and the maximum step to change the size of a parking lot in a given bus, respectively.

$$Cost^{INV} = C^{PL} \sum_{b \in B} Size_b^{PL}, B = \{1, \ldots, N^B\} \tag{6.19}$$

$$Size_b^{PL} = n_b^{PL} \times Size^{PL-MIN}, n_b^{PL} \in \{1, \ldots, n^{PL-MAX}\}, \forall b \in B \tag{6.20}$$

6.5.2.2 Maintenance Cost

Equation (6.21) shows the present worth value of maintenance cost of parking lots ($) installed in the electrical distribution system during the given planning period. As can be seen, a certain percentage of investment cost (μ) is defined as the yearly maintenance cost of parking lots. Herein, y, Y, N^Y, IFR, and ITR are the index of year, the set of years of planning period, the length of planning time horizon (in years), the inflation rate (percent), and the interest rate (percent), respectively.

$$\begin{aligned} PWV\left(Cost^M\right) &= \sum_{y \in Y} \left(\frac{\mu}{100} \times Cost^{INV}\right) \times \left(\frac{1 + IFR/100}{1 + ITR/100}\right)^y, Y \\ &= \{1, \ldots, N^Y\} \end{aligned} \tag{6.21}$$

6.5.2.3 Incentive Cost

The present worth value of incentives ($) paid to the drivers to motivate them to provide the V2G service during the planning period is given in (6.22). As can be seen, it has a dependency on the amount of incentive (percent) offered to the responsive drivers (β) to provide the V2G service, the electricity price (π^E) in $/MWh, the expected V2G power of a PEV ($\widetilde{P^{V2G}}$) in MW, and the inflation and interest rates (percent).

As can be seen in (6.23), the expected V2G power of a PEV is the function of responsiveness probability of driver (ξ^{SC} in percent), the driver's social class (*SC*), the value of incentive (β) in percent, the driver's distance from the parking lot (α) in km, the rated power of PEV's battery (P^{PEV}) in kW, the efficiency of PEV's battery for V2G power provision (Eff^{PEV}) in percent, the state of charge (SOC) of PEV (percent), and the defined depth of discharge (DOD) in percent. Herein, *d*, *D*, *t*, *T*, *e*, *PEVs*, N^{PEVs}, *LI*, *MI*, and *HI* are the index of day, the set of days of a year, the index of time, the set of hours of a day, the index of PEV, the set of PEVs, the total number of PEVs in the area, the index of a low-income driver, the index of a moderate-income driver, and the index of a high-income driver, respectively.

$$\begin{aligned} PWV\left(Cost^{INC}\right) &= \sum_{y \in Y}\left(\sum_{d \in D}\sum_{t \in T}\sum_{e \in PEVs}\frac{\beta_t}{100} \times \pi_t^E \times \widetilde{P_{e,t}^{V2G}}\right) \\ &\quad \times \left(\frac{1 + IFR/100}{1 + ITR/100}\right)^y, D \\ &= \{1, \ldots, 365\}, T = \{1, \ldots, 24\}, PEVs \\ &= \{1, \ldots, N^{PEVs}\}, 0 \le \beta \le 100 \end{aligned} \tag{6.22}$$

$$\begin{aligned} \widetilde{P_{e,t}^{V2G}} = \frac{\xi_{e,\alpha,\beta}^{SC}}{100} \times \frac{P_e^{PEV}}{1000} \times \frac{Eff_e^{PEV}}{100} \times \left(\frac{SOC_{e,t}^{PEV} - DOD^{PEV}}{100}\right), \\ \forall t \in T, \forall e \in PEVs, SC \in \{LI, MI, HI\}, 0 \le \beta \le 100, 0 \le \alpha \le 10\ \text{km} \end{aligned} \tag{6.23}$$

6.5.2.4 Energy Loss Cost

The amount of energy which is lost in the feeder's branches (MWh) and the present worth value of energy loss cost during the planning time horizon ($) are presented in (6.24) and (6.25), respectively. Herein, *R* and $|I|$ are the resistance of a branch (ohm) and the magnitude of current flowing through the branch (ampere), respectively.

$$E^{LOSS} = 10^{-6} \sum_{y \in Y} \sum_{d \in D} \sum_{t \in T} \sum_{b \in B} R_b \times |I_{b,t}|^2 \tag{6.24}$$

$$PWV\left(Cost^{LOSS}\right) = 10^{-6} \sum_{y \in Y} \left(\sum_{d \in D} \sum_{t \in T} \sum_{b \in B} R_b \times |I_{b,t}|^2 \times \pi_t^E \right) \times \left(\frac{1 + IFR/100}{1 + ITR/100} \right)^y \tag{6.25}$$

6.5.2.5 ENS Cost

Equation (6.26), as the extended version of (6.15), presents the amount of ENS (MWh) due to the faults and failures occurred in the feeder's branches. Moreover, Eq. (6.27) gives the present worth value of ENS cost during the planning period ($). Herein, λ^{Model}, T^{FL}, LNS^{FL}, π^{CD}, T^{FR}, and LNS^{FR} are the failure rate of a branch of feeder (f/y), the failure locating duration (hour), the load not supplied during the failure locating process (MW), the customer's damage due to the power outage ($/MWh), the failure repairing duration (hour), and the load not supplied during the failure repairing process (MW), respectively. In addition, *Model*, *CON*, *LIN*, *POW*, *EXP*, and *LOG* are the set of FFR models and the indices of constant, linear, power, exponential, and logarithmic FFR models, respectively. Moreover, *c*, *C*, *Res*, *Com*, and *Ind* are the index of a customer's type, the set of customers' types, and the indices of residential, commercial, and industrial consumers, respectively.

$$ENS = \sum_{y \in Y} \sum_{b \in B} \lambda_b^{Model} \times \left(T^{FL} \sum_{i \in J} LNS_i^{FL} + T^{FR} \sum_{i \in J} LNS_i^{FR} \right), Model \in \{CON, LIN, POW, EXP, LOG\} \tag{6.26}$$

$$PWV\left(Cost^{ENS}\right) = \sum_{y \in Y} \left(\sum_{t \in T} \sum_{b \in B} \lambda_b^{Model} \times \left[T^{FL} \sum_{i \in J} \sum_{c \in C} LNS_{i,c}^{FL} \times \pi_{c,t}^{CD} + T^{FR} \sum_{i \in J} \sum_{c \in C} LNS_{i,c}^{FR} \times \pi_{c,t}^{CD} \right] \right) \times \left(\frac{1+IFR/100}{1+ITR/100} \right)^y, C \in \{Res, Com, Ind\} \tag{6.27}$$

6.5.3 *Problem Constraints*

6.5.3.1 Power Flow of Branches

One of the security constraints of system is concerned with the magnitude limit of apparent power flowing through each branch of feeder, presented in (6.28). Herein, $|MVA|$ and TC are the magnitude of apparent power flowing through the branch and the thermal capacity of branch, respectively.

$$|MVA_b| \leq TC_b, \forall b \in B \tag{6.28}$$

6.5.3.2 Voltage Profile of Buses

Another security constraint of system is the voltage magnitude bounds of each bus, as can be seen in (6.29). Herein, $min|V|$, $|V|$, and $max|V|$ are the lower limit for the voltage magnitude of bus, the magnitude of operating voltage of bus, and the upper limit for the voltage magnitude of bus, respectively.

$$min|V| \leq |V_i| \leq max|V|, \forall i \in J \tag{6.29}$$

6.6 Optimization Technique

In this chapter, genetic algorithm (GA) is applied to solve the optimization problem [16]. Herein, every bus of feeder is considered as a candidate to install a parking lot. However, the size of a parking lot must be within the minimum size ($Size^{PL-MIN}$) and maximum size ($n^{PL-MAX} \times Size^{PL-MIN}$). In this study, the value of $Size^{PL-MIN}$ and n^{PL-MAX} are considered about 50 and 3, respectively. Thus, to code the value of n^{PL} (as one of the problem variables) in GA, two bits (two genes) are needed, where 00, 01, 10, and 11 are equivalent to 0, 1, 2, and 3 (or a parking lot with the size of 0, 50, 100, and 150 PEVs), respectively.

In addition, the value of incentive (incentive) introduced to the drivers is another type of problem variable. Therefore, to model the hourly value of incentive in GA, four bits are needed. Herein, 0000, 0001, 0010, 0011, 0100, 0101, 0110, 0111, 1000, 1001, 1010 are equivalent to the sequence of 0, 1, 2, …, 10 or 0%, 10%, 20%, …, 100% incentive, respectively.

In this study, to optimally solve the planning problem, each parking lot, at each hour of typical day, may be closed to prevent the PEVs from the V2G service provision. Therefore, "1" and "0" mean that the parking lot is open and close at that hour, respectively. Figure 6.11 illustrates the structure of a chromosome designed for this study. As can be seen, it is a matrix with the size of $24 \times (2N^B + 4)$, where 24, 2, N^B, and 4 are the number of hours of a day, the number of genes needed to model n^{PL}, the number buses of feeder, and the number of genes needed to model the hourly value of incentive, respectively.

The inverse value of total cost of planning problem during the planning period is defined as the fitness of each chromosome, and then GA tries to improve the fitness of chromosomes of population. In the following, the steps to apply GA in the planning problem are presented and described. In addition, Fig. 6.12 illustrates the flowchart to apply GA in the planning problem.

Step 1: Obtaining the primary data
The parameters of GA: These parameters include the mutation probability of a gen (θ^M) and the number of chromosomes as the size of population (N^{GA}).

The parameters of problem and system: The value of all the parameters, presented in Sect. 6.7.1, are obtained.

The primary population: The chromosomes of population are randomly quantified with the binary digits.

Step 2: Updating the population
Applying mutation operator: The mutation operator, with the occurrence probability of θ^M, is applied on each gene of every chromosome of population, as can be seen in Fig. 6.13.

Applying crossover operator: Proportional to the number of columns of a chromosome ($2N^B + 4$), the crossover points are randomly selected for each two chromosomes, and then the crossover operator is applied on them to reproduce two new offspring, as can be seen in Fig. 6.14.

Step 3: Selecting the new population
Evaluating the fitness of each chromosome: For every chromosome of population, the planning problem is solved, and if all the constraints are satisfied, the fitness of chromosomes is calculated.

Applying the selection process: A probabilistic fitness-based selection process is applied on the old chromosomes of population to select the new generation of chromosomes. In the selection process, the fitter chromosomes are more likely to be chosen, since the selection probability of a chromosome is proportional to its fitness, as can be seen in (6.30). Herein f_{ch} and $Rand[0, 1)$ are the fitness of a chromosome and a random number in the range of [0,1), respectively. In addition, $max\{f_1, \ldots, f_{ch}, \ldots, f_{N^{GA}}\}$ determines the maximum fitness of chromosomes of population.

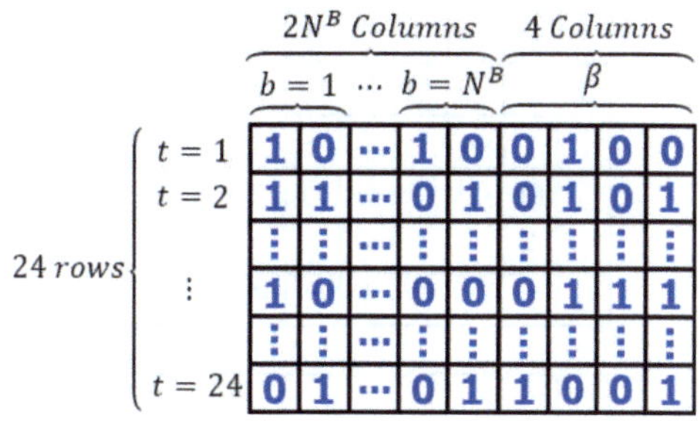

Fig. 6.11 The structure of a chromosome

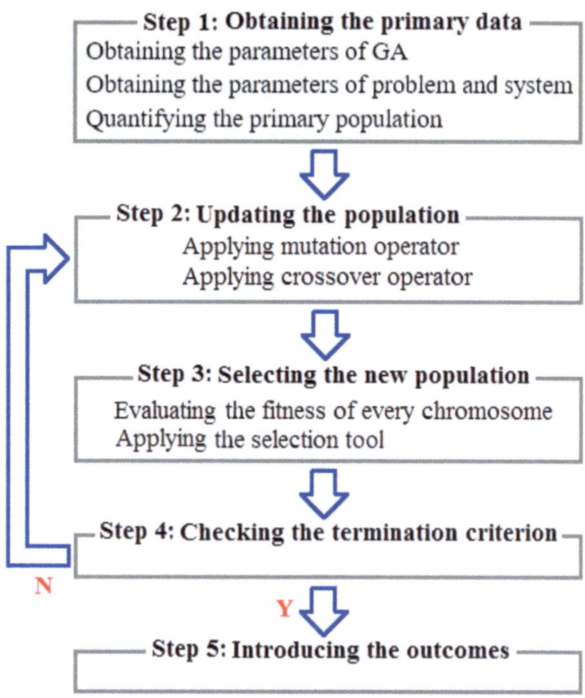

Fig. 6.12 The flowchart for applying GA in the planning problem

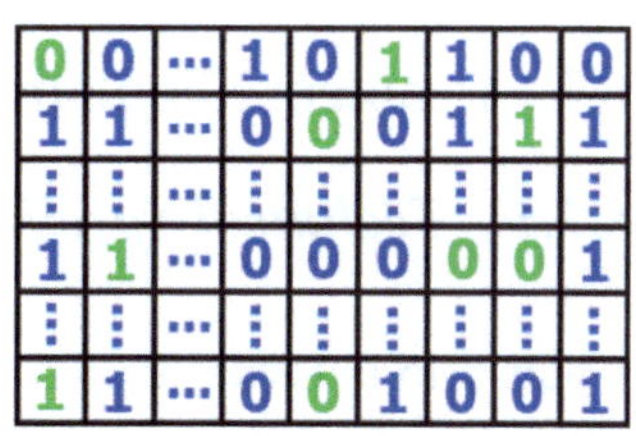

Fig. 6.13 Applying mutation operator on the genes of a chromosome

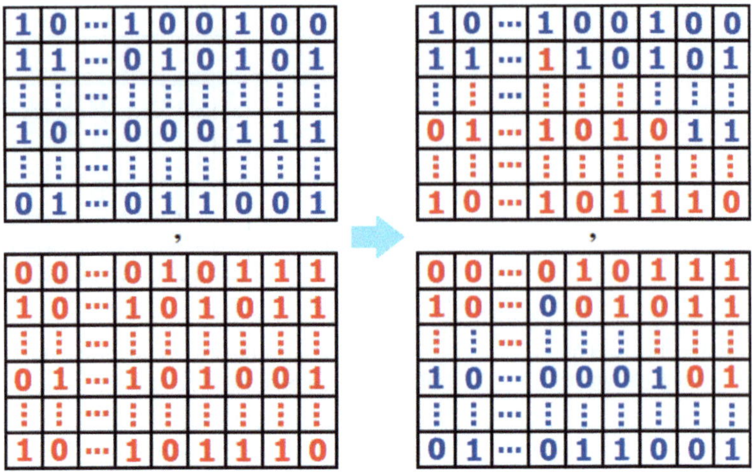

Fig. 6.14 Applying crossover operator on two chromosomes to reproduce the new chromosomes

$$Selection = \begin{cases} Yes & \frac{f_{ch}}{max\{f_1,\ldots,f_{ch},\ldots,f_{NGA}\}} > Rand[0,1) \\ No & Otherwise \end{cases} \quad (6.30)$$

Step 4: Checking the termination criterion
In this stage, the convergence status of optimization procedure is checked. Based on this, if no remarkable improvement is identified in the fitness of new chromosomes (compared to the fitness of old chromosomes), the optimization process is finalized; otherwise, the algorithm is continued form Step 2.

Step 5: Introducing the outcomes
The consequences include the optimal value of problem variables, that is, the optimal hourly location and size of parking lots and the optimal hourly value of incentive.

6.7 Problem Simulation

6.7.1 Technical Specifications of System and Problem

Figure 6.15 illustrates the single-line diagram of electrical distribution system under study and the hourly geographical routes of PEVs in San Francisco, CA, USA [14]. As can be seen, the distribution network includes two medium-voltage electrical distribution feeders (F1 and F2) supplied by the 33/11 kV sub-transmission transformer. Each feeder includes a circuit breaker (C.B.) to discount the feeder from the transmission system due to the occurrence of a permanent fault in the distribution

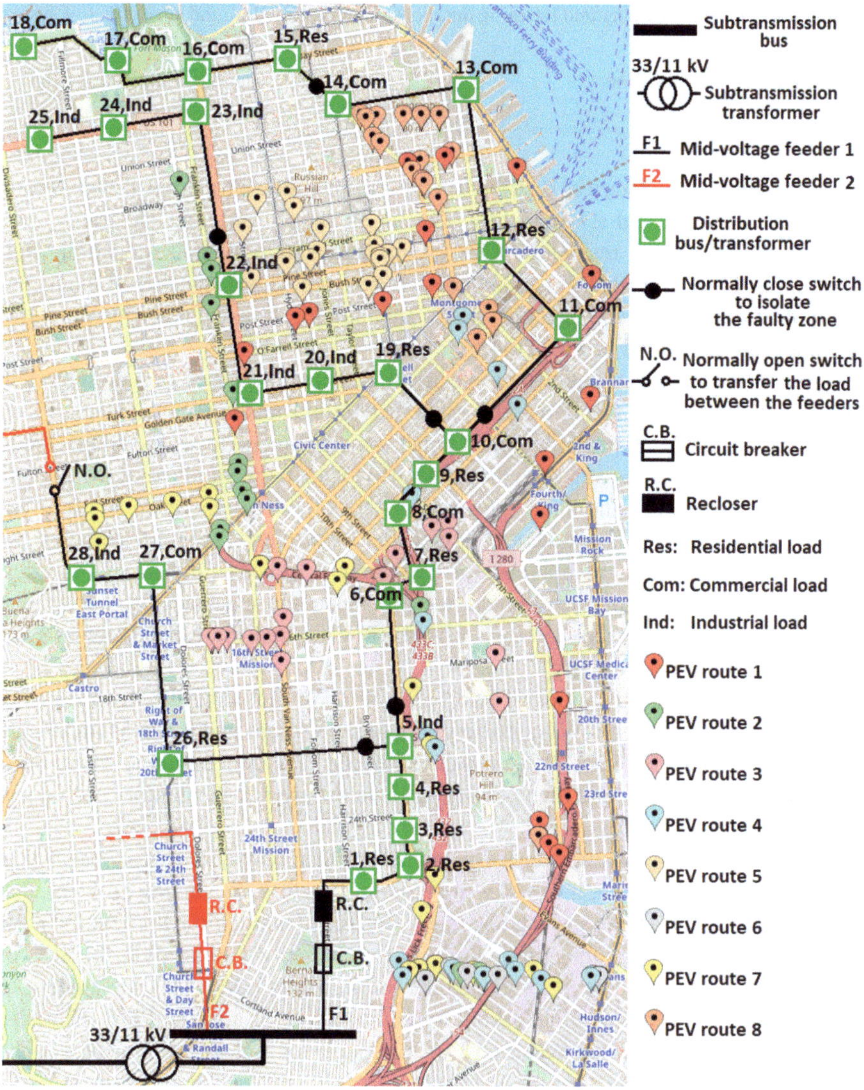

Fig. 6.15 The single-line diagram of electrical distribution system under study and the hourly geographical routes of PEVs in San Francisco, CA, USA

system. The C.B. will not be closed until the fault is permanently removed. In addition, they include a recloser (R.C.) to automatically disconnect, and then connect the feeder to the sub-transmission bus if the failure is temporary. In this chapter, the first feeder (F1) is studied, and the second one (F2) is utilized to transfer part of load of F1 to F2 during the fault occurrence using the normally open switch. Herein, the probability of transferring is about 60%.

Table 6.4 The fault locating and repairing durations (hour) required for the assessment of system reliability

Fault locating duration (hour)	Fault repairing duration (hour)
1	5

In addition, the electrical distribution system includes 28 electrical distribution buses/transformers and 27 branches. The buses supply different electricity consumers including the residential (Res), commercial (Com), and industrial (Ind) customers. Moreover, the system includes six normally close switches in the beginning of branches to isolate the faulty zone. The fault locating and repairing durations (in hour) are presented in Table 6.4.

The technical specifications of electrical distribution buses are presented in Table 6.5. It presents the ID of buses, the active and reactive power demands of buses (MW, MVAr) at peak hour (21st hour), the consumers' load type connected to each bus, the rated power of a customer (kW), and the number of consumers connected to the bus. The hourly demand pattern of system and each bus (p.u.) in the typical day is demonstrated in Fig. 6.16. Moreover, Table 6.6 presents the technical specifications of branches of F1, where it indicates the ID of branches, the related two buses of each branch, the resistance and reactance of each branch (p.u.), the thermal capacity of each branch (MVA), and the failure rate of each branch in failure per year (f/y).

The technical specifications of two types of PEV, including the rated capacity (kWh) and rated power (kW) of battery of Citroën C-Zero and Tesla Model S, are presented in Table 6.7 [22]. Table 6.8 classifies the PEV penetration levels in three groups (low, moderate, and high penetration levels) based on the number of PEVs in the area and their driving routes.

The value of other parameters of planning problem and system are presented in Table 6.9. As can be seen, it includes the economic factors of planning problem, that is, the inflation rate (*IFR*) and interest rate (*ITR*), the planning time horizon (N^Y), the lower and upper voltage magnitude limits of buses ($min|V|$ and $max|V|$), the SOC of PEVs (SOC^{PEV}), the value of DOD defined for the PEVs (DOD^{PEV}), the amount of investment needed to install a parking lot and equip it for a single PEV (C^{PL}), the parameter of yearly maintenance cost of a parking lot and its equipment (μ), the maximum step to change the size of a parking lot (n^{PL-MAX}), the minimum size of a parking lot ($Size^{PL-MIN}$), the efficiency of PEV's battery for the V2G service provision (Eff^{PEV}), the mutation probability of a gene in GA (θ^M), and the number of chromosomes of population in GA (N^{GA}).

Figure 6.17 illustrates the hourly electricity price ($/MWh) in the typical day for different electricity consumer types. In addition, the hourly damage ($/MWh) imposed to different consumers in the typical day due to the power outage is graphically shown in Fig. 6.18. As can be seen, different electricity consumers are dissimilarly affected by the power interruption, where the industrial and residential consumers are imposed the highest and lowest damages, respectively.

Table 6.5 The technical specifications of electrical distribution buses of first feeder (F1)

Bus no.	Active power demand (MW) at peak hour (21st hour)	Reactive power demand (MVAr) at peak hour (21st hour)	Load type	Rated power of a customer (kW)	No. of customers connected to the bus
1	0.22	0.10	Residential (Res)	15	15
2	0.22	0.10	Residential (Res)	15	15
3	0.55	0.25	Residential (Res)	15	37
4	0.55	0.25	Residential (Res)	15	37
5	0.60	0.26	Industrial (Ind)	150	4
6	0.60	0.26	Commercial (Com)	30	20
7	0.60	0.25	Residential (Res)	15	40
8	0.65	0.28	Commercial (Com)	30	22
9	0.65	0.28	Residential (Res)	15	43
10	0.70	0.30	Commercial (Com)	30	23
11	0.75	0.31	Commercial (Com)	30	25
12	0.75	0.32	Residential (Res)	15	50
13	0.75	0.31	Commercial (Com)	30	25
14	0.55	0.25	Commercial (Com)	30	18
15	0.95	0.46	Commercial (Com)	30	63
16	0.95	0.43	Residential (Res)	15	32
17	0.95	0.44	Commercial (Com)	30	32
18	0.95	0.45	Commercial (Com)	30	32
19	0.95	0.46	Residential (Res)	15	63
20	0.95	0.46	Industrial (Ind)	150	6
21	0.95	0.47	Industrial (Ind)	150	6
22	1.00	0.49	Industrial (Ind)	150	7
23	1.00	0.40	Industrial (Ind)	150	7
24	1.00	0.40	Industrial (Ind)	150	7
25	1.05	0.41	Industrial (Ind)	150	7
26	0.95	0.40	Residential (Res)	15	63
27	0.95	0.40	Commercial (Com)	30	32
28	0.95	0.40	Industrial (Ind)	150	6

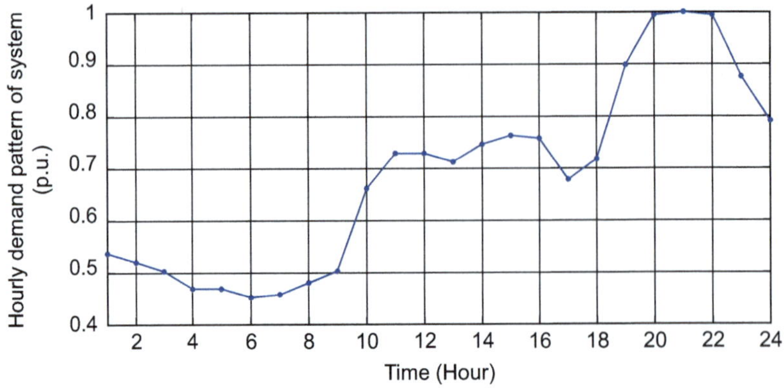

Fig. 6.16 The hourly demand pattern of system and each bus of system (p.u.) in the typical day [17]

Table 6.6 The technical specifications of branches of first feeder (F1)

ID of branch	Related buses of branch	Resistance (p.u.)	Reactance (p.u.)	Thermal capacity (MVA)	Failure rate (f/y)
1	1–2	0.0020	0.0010	25	9
2	2–3	0.0037	0.0054	25	9
3	3–4	0.0044	0.0040	25	9
4	4–5	0.0048	0.0042	25	9
5	5–6	0.0039	0.0049	18	9
6	6–7	0.0040	0.0065	18	9
7	7–8	0.0050	0.0051	18	9
8	8–9	0.0049	0.0056	18	9
9	9–10	0.0053	0.0056	18	9
10	10–11	0.0043	0.0014	8	9
11	11–12	0.0046	0.0027	8	9
12	12–13	0.0040	0.0042	8	9
13	13–14	0.0048	0.0050	8	9
14	14–15	0.0059	0.0044	8	9
15	15–16	0.0057	0.0049	8	9
16	16–17	0.0036	0.0025	8	9
17	17–18	0.0036	0.0025	8	9
18	10–19	0.0035	0.0013	8	9
19	19–20	0.0047	0.0015	8	9
20	20–21	0.0054	0.0034	8	9
21	21–22	0.0049	0.0029	8	9
22	22–23	0.0063	0.0032	8	9
23	23–24	0.0056	0.0014	8	9
24	24–25	0.0055	0.0013	8	9
25	5–26	0.0063	0.0032	8	9
26	26–27	0.0056	0.0014	8	9
27	27–28	0.0055	0.0013	8	9

Table 6.7 The technical specifications of two PEV types [22]

–	Citroën C-Zero	Tesla Model S
Battery capacity (kWh)	16	60
Battery power (kW)	49	225

Table 6.8 Classifying the PEV penetration levels based on the number of PEVs in the area and their driving routes

–		PEV penetration level		
		Low	Moderate	High
Number of PEVs with different driving routes	Driving route 1	125	250	375
	⋮	⋮	⋮	⋮
	Driving route 8	125	250	375
Total number of PEVs in the area (N^{PEVs})		1000	2000	3000

Table 6.9 The value of parameters of planning problem and system

Parameter	Value	Parameter	Value
IFR	10%	C^{PL}	\$1000
ITR	15%	μ	10%
N^{Y}	20 years	n^{PL-MAX}	3
$min\|V\|$	0.90 p.u.	$Size^{PL-MIN}$	50
$max\|V\|$	1.05 p.u.	Eff^{PEV}	95%
SOC^{PEV}	100%	θ^{M}	10%
DOD^{PEV}	20%	N^{GA}	100

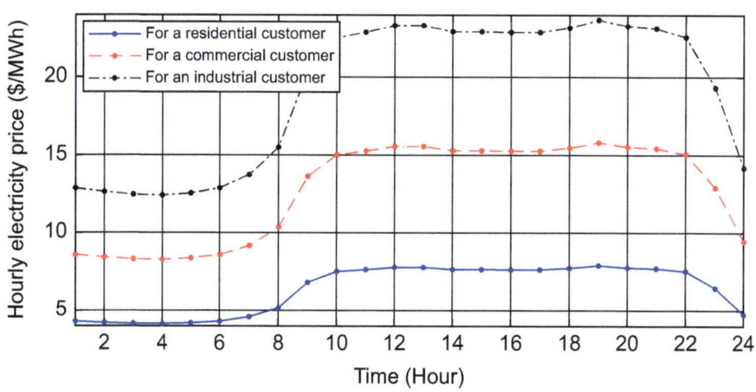

Fig. 6.17 The hourly electricity price (\$/MWh) in the typical day for different customers

Table 6.10 presents the description of different scenarios defined in this chapter. As can be seen, ten scenarios are defined to study the planning problem before and after the parking lot placement and sizing. In all the scenarios, the FFR (linear

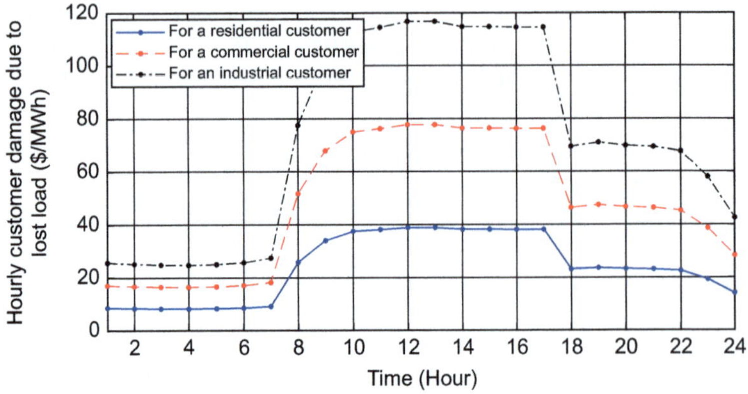

Fig. 6.18 The hourly customer's damage due to the power outage ($/MWh) in the typical day for different customers

Table 6.10 The description of different scenarios under study

–	Sc. no.	PEV type	Social class of drivers	PEV pen.	FFR modelling	VDL modelling	Description
Before planning	1	Citroën C-Zero	–	–	Yes	Yes	Studying the primary condition of system
After planning	2	Citroën C-Zero	LI	Low	Yes	Yes	Studying the effects of social class of drivers
	3		MI		Yes	Yes	
	4		HI		Yes	Yes	
	5	Citroën C-Zero	MI	Low	Yes	Yes	Studying the effects of PEV penetration level
	6			Moderate	Yes	Yes	
	7			High	Yes	Yes	
	8	Tesla Model S	MI	Low	Yes	Yes	Studying the effects of PEV type
	9	Citroën C-Zero	MI	Low	No	Yes	Studying the effects of FFR (linear model)
	10	Citroën C-Zero	MI	Low	Yes	No	Studying the effects of VDL

model) and VDL are modelled unless it is indicated. The other models of FFR, that is, power, exponential, and logarithmic models, are not studied in this chapter, and therefore they are considered as the future study.

In the first scenario, no parking lot is allocated to the electrical distribution system, and the primary energy loss during the planning period, the value of ENS during the planning period, the value of reliability indices in the typical year, the voltage profile of buses at peak hour of typical day, the amount of power flowing through the branches at peak hour of typical day, and the total cost of problem during the planning period are calculated. The reliability indices include SAIFI in failure/year (f/y) per customer, SAIDI in hour/year per customer, ENS in MWh/year, AENS in MWh/year per customer, and ASAI in percent.

In the second-fourth scenarios, the effects of social class of drivers on the planning problem are investigated. Herein, the parking lot placement and sizing problem is solved for each social class of drivers (low-income, moderate-income, and high-income) considering Citroën C-Zero PEVs and low PEV penetration level. In these scenarios, in addition to the abovementioned outputs, the optimal hourly value of incentive and the optimal hourly size and location of parking lots in the typical day are determined.

In the fifth-seventh scenarios, the effects of PEV penetration level are studied. Herein, the planning problem is solved for each PEV penetration level (low, moderate, and high) considering Citroën C-Zero PEVs and moderate-income drivers.

In the eighth scenario, the type of PEVs is changed to Tesla Model S to study the impacts of PEV type on the problem consequences considering moderate-income drivers and low PEV penetration level.

In the ninth and tenth scenarios, the effects of FFR (linear model) and VDL modelling on the problem results are analyzed considering moderate-income drivers and low PEV penetration level, respectively.

6.7.2 Studying the Primary Condition of System

In this part, the first scenario is studied. In this scenario, no parking lot is placed in the electrical distribution system; however, FFR (linear model) and VDL are modelled. Table 6.11 presents the simulation results of planning problem before parking lot allocation. As can be seen, the total energy loss, total ENS, and total cost of problem during the planning period are about 167.965 MWh, 225.188 MWh, and $9,270,118, respectively.

Moreover, the primary voltage profile of buses (p.u.) at peak hour (21st hour) of typical day is shown in Fig. 6.19. As can be seen, the voltage profile of several buses (bus IDs: 10–25) have violated, since they are below 0.9 p.u. In addition, Fig. 6.20 illustrates the thermal capacity of branches of feeder (MVA) and the apparent power flowing through the branches (MVA) at peak hour (21st hour) of typical day. As is shown, the thermal capacity constraints of branches with the IDs of 1, 2, 5, and 6 have violated, since extra apparent power is flowing through them. Therefore, the security condition of system is not satisfactory.

Table 6.11 The results of planning problem simulation before parking lot allocation

Outputs	Value
Energy loss during the planning period (GWh)	167.965
ENS during the planning period (GWh)	225.188
SAIFI (f/y per customer) in the typical year	243
SAIDI (h/y per customer) in the typical year	723
ENS (MWh/y) in the typical year	11,259
AENS (MWh/y per customer) in the typical year	15.2
ASAI (%) in the typical year	91.7
Incentive cost during the planning period ($)	0
Investment cost during the planning period ($)	0
Maintenance cost during the planning period ($)	0
Energy loss cost during the planning period ($)	1,580,905
ENS cost during the planning period ($)	7,689,213
Total cost of problem during the planning period ($)	9,270,118

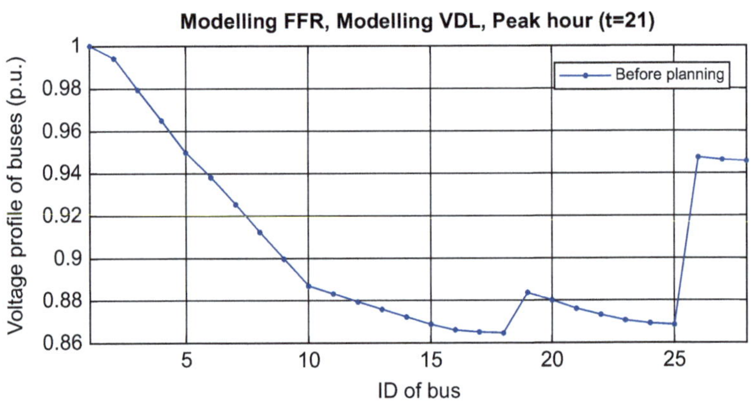

Fig. 6.19 The primary voltage profile of buses (p.u.) at peak hour (21st hour) of typical day modelling FFR (linear) and VDL

6.7.3 *Studying the Effects of Social Class of Drivers*

In this section, the second-fourth scenarios are investigated, where the effects of social class of drivers (low-income, moderate-income, and high-income) on the planning problem are studied. In this part, the type of PEVs and PEV penetration level are Citroën C-Zero and low, respectively, and FFR (linear model) and VDL are modelled.

The optimal hourly location and size of parking lots in the typical day for low-income, moderate-income, and high-income drivers are illustrated in Figs. 6.21, 6.22, and 6.23, respectively. As can be observed, different sizes of parking lots are

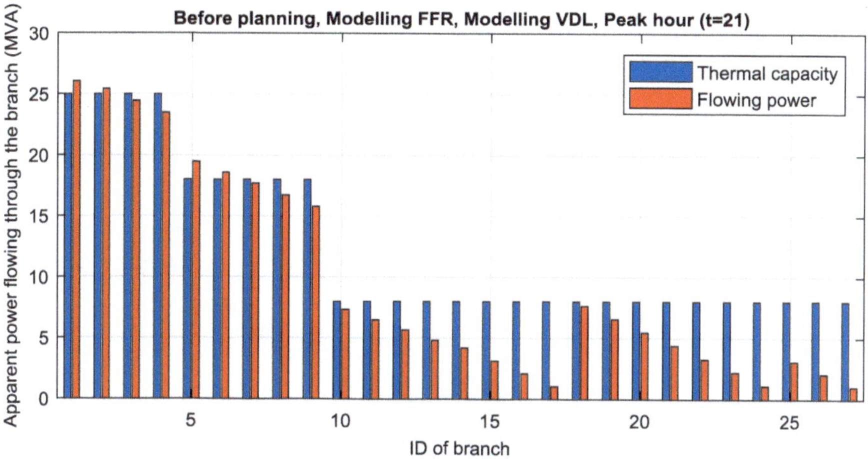

Fig. 6.20 The thermal capacity of branches of feeder (MVA) and apparent power flowing through the branches (MVA) at peak hour (21st hour) of typical day before solving the planning problem, modelling FFR (linear), and VDL

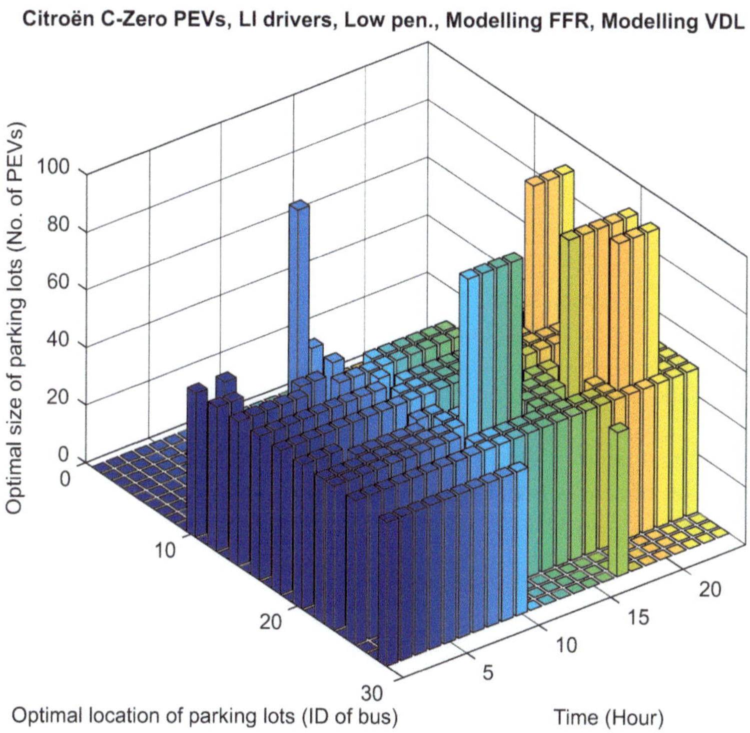

Fig. 6.21 The optimal hourly location and size of parking lots in the typical day for low-income drivers considering Citroën C-Zero PEVs and low PEV penetration level and modelling FFR (linear) and VDL

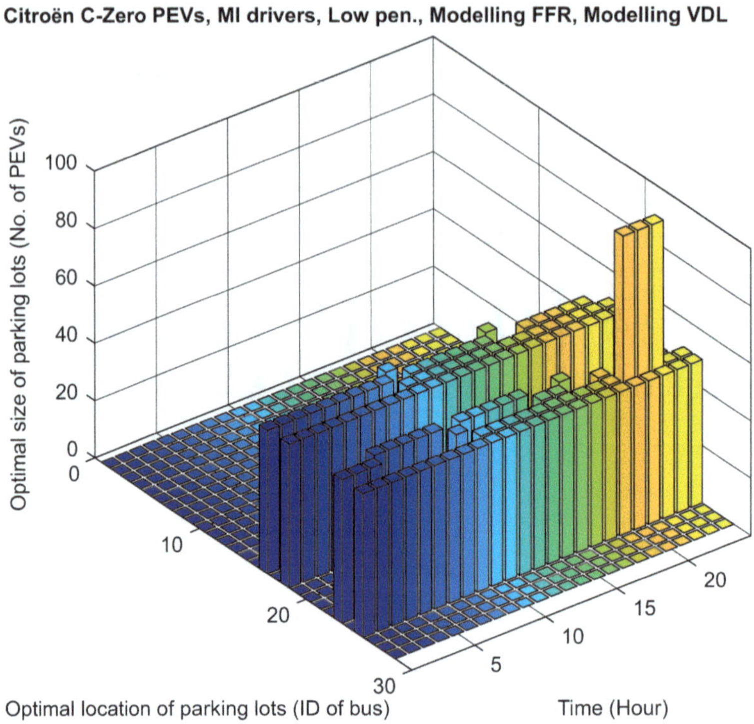

Fig. 6.22 The optimal hourly location and size of parking lots in the typical day for moderate-income drivers considering Citroën C-Zero PEVs and low PEV penetration level and modelling FFR (linear) and VDL

allocated to some certain buses. Moreover, the parking lots need to be closed during some periods of typical day. In addition, the most and the largest parking lots are placed in the network if the social class of drivers is low-income; however, the least and the smallest parking lots are allocated to the system for the case of high-income drivers.

The optimal hourly value of incentive (percent) in the typical day for different social classes of drivers is shown in Fig. 6.24. As can be seen, the optimal value of incentive needs to be updated during the day. Moreover, the largest and smallest incentives are needed to motivate the high-income and low-income drivers to participate in the V2G service provision.

Figure 6.25 illustrates the sensitivity analysis for the planning problem cost ($) at peak hour (21st hour) of typical day with respect to the value of incentive (%) for different social classes of drivers. As can be seen, for each social class of drivers, there is just one specific point on the curve corresponding to the minimum cost and optimal incentive. In addition, the optimal point concerned with each social class is distinctive. Moreover, the participation of low-income and high-income drivers in the V2G service provision result in the least and the most values of minimum cost, respectively.

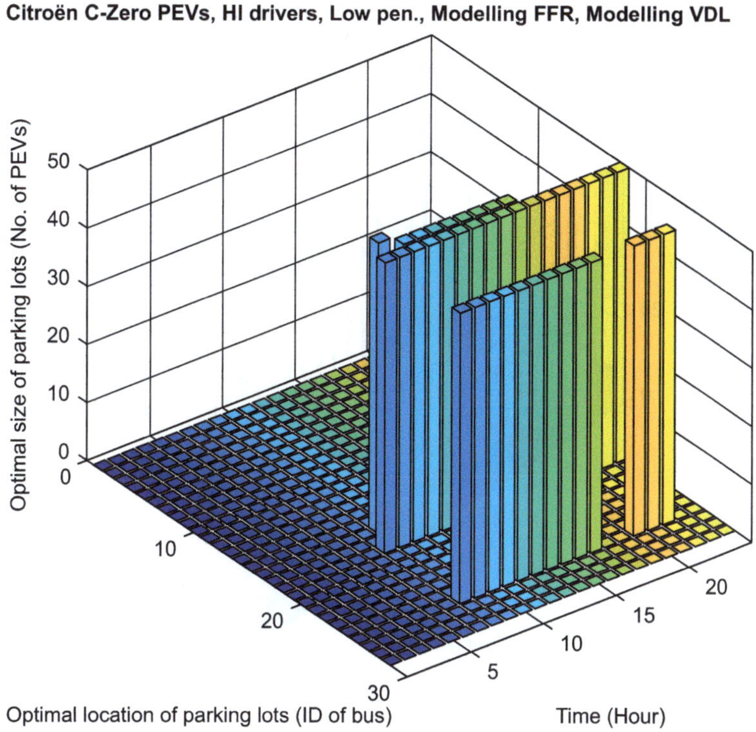

Fig. 6.23 The optimal hourly location and size of parking lots in the typical day for high-income drivers considering Citroën C-Zero PEVs and low PEV penetration level and modelling FFR (linear) and VDL

Figure 6.26 shows the primary and final hourly energy loss of system in the typical day (MWh) for different social classes of drivers. As can be seen, after optimal parking lot sizing and placement, the hourly energy loss of system dramatically decreases, specifically for the low-income drivers. However, the energy loss of system does not change at some hours of typical day in the case of high-income drivers.

The primary and final energy loss and ENS of system during the planning period (GWh) for different social classes of drivers are displayed in Fig. 6.27. As can be seen, both parameters have the most desirable value for the participation of low-income drivers in the V2G service provision, where there is a reduction of 70% and 78% for the energy loss and ENS during the planning period, respectively.

Figure 6.28 shows the primary and final reliability indices of system in the typical year for different social classes of drivers. As can be seen, all the reliability indices are improved due to the optimal parking lot allocation and sizing. Herein, the simulation of planning problem for the low-income and high-income drivers results in the minimum and maximum values of SAIFI, SAIDI, and AENS (as the

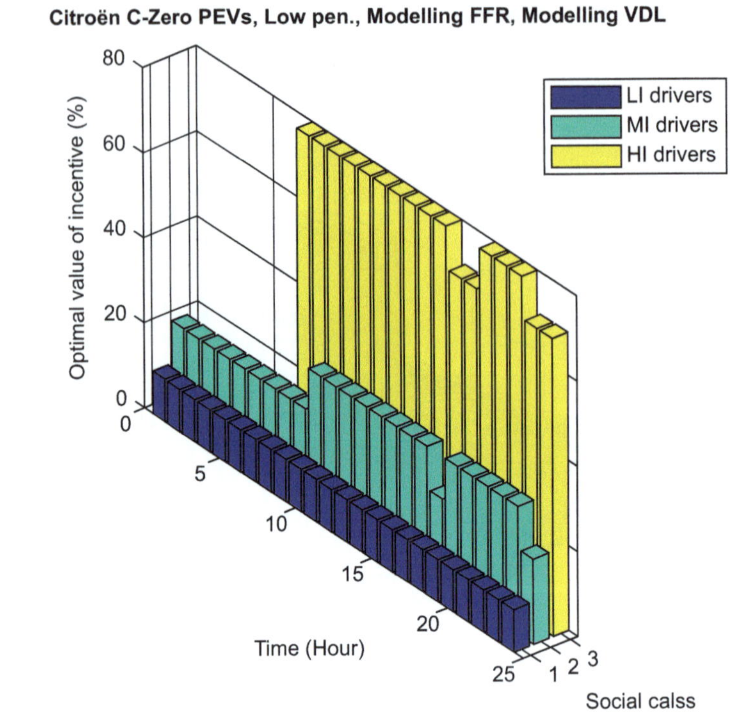

Fig. 6.24 The optimal hourly value of incentive (%) in the typical day for different social classes of drivers considering Citroën C-Zero PEVs and low PEV penetration level and modelling FFR (linear) and VDL

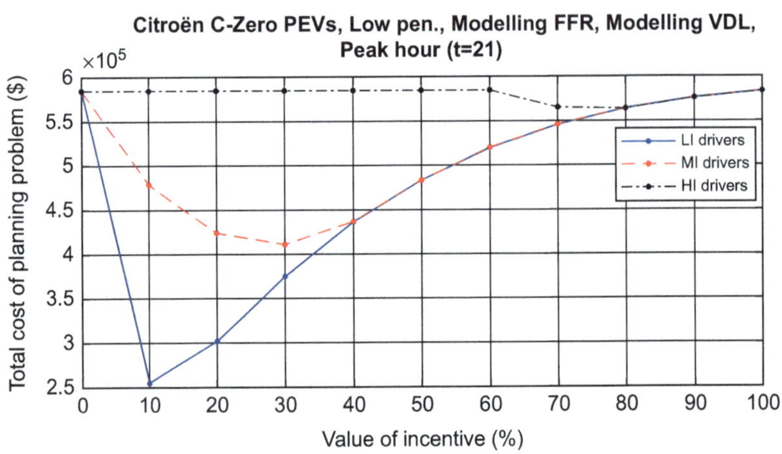

Fig. 6.25 The sensitivity analysis for the planning problem cost ($) at peak hour (21st hour) of typical day with respect to the value of incentive (%) for different social classes of drivers considering Citroën C-Zero PEVs and low PEV penetration level and modelling FFR (linear) and VDL

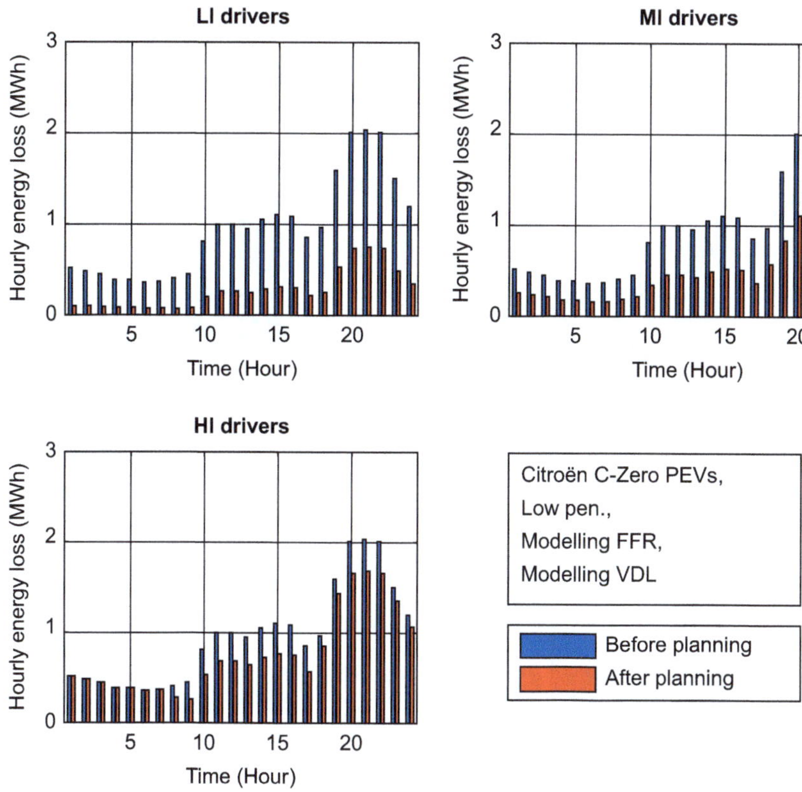

Fig. 6.26 The primary and final hourly energy loss of system in the typical day (MWh) for different social classes of drivers considering Citroën C-Zero PEVs and low PEV penetration level and modelling FFR (linear) and VDL

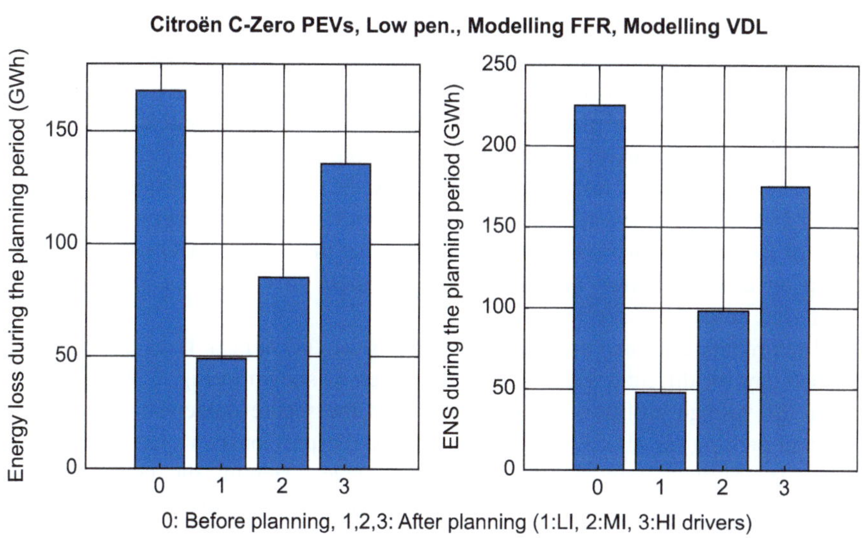

Fig. 6.27 The primary and final energy loss and ENS of system during the planning period (GWh) for different social classes of drivers considering Citroën C-Zero PEVs and low PEV penetration level and modelling FFR (linear) and VDL

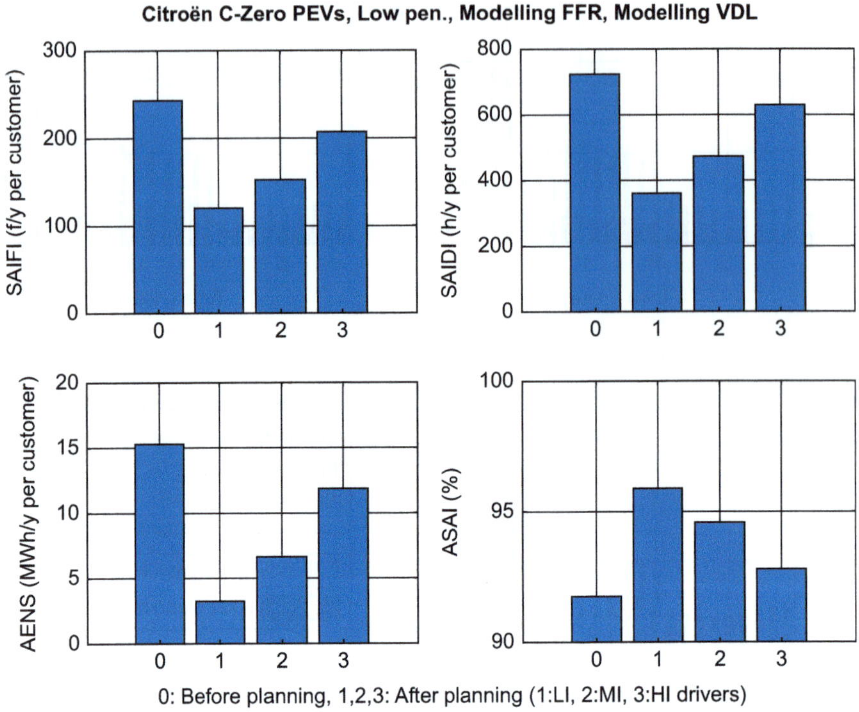

Fig. 6.28 The primary and final reliability indices of system in the typical year for different social classes of drivers considering Citroën C-Zero PEVs and low PEV penetration level and modelling FFR (linear) and VDL

undesirable indices) and the maximum and minimum value of ASAI (as the favorable index), respectively.

Figure 6.29 illustrates the primary and final cost terms and total cost of problem during the planning period ($) for different social classes of drivers. As can be noticed, the total cost of planning problem reduces by the cooperation of any social class of drivers; however, the low-income and high-income drivers have the most and least contributions, respectively. Regarding the cost terms of planning problem, it is noticed that the investment and maintenance costs have the largest values in the case of low-income drivers. This fact implies that the local DISCO needs to invest more on the installation of parking lots for these drivers, since they are more interested in the V2G service provision.

The primary and final voltage profile of buses (p.u.) at peak hour (21st hour) of typical day for different social classes of drivers is shown in Fig. 6.30. As can be seen, the voltage profile of buses is placed in the secure range (0.90–1.05) after optimal parking lot allocation. In other words, the previously violated buses (bus IDs: 10–25) have now a standard voltage level. Herein, the cooperation of low-income drivers results in the best voltage profile for the system.

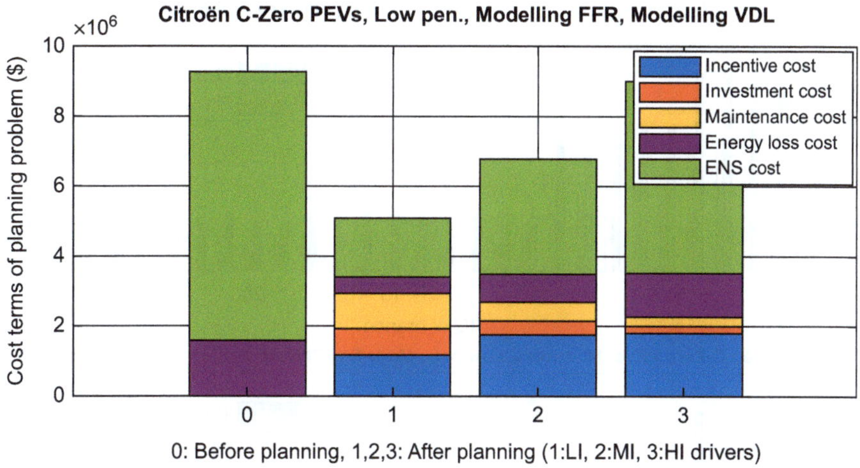

Fig. 6.29 The primary and final total cost and cost terms of problem during the planning period ($) for different social classes of drivers considering Citroën C-Zero PEVs and low PEV penetration level and modelling FFR (linear) and VDL

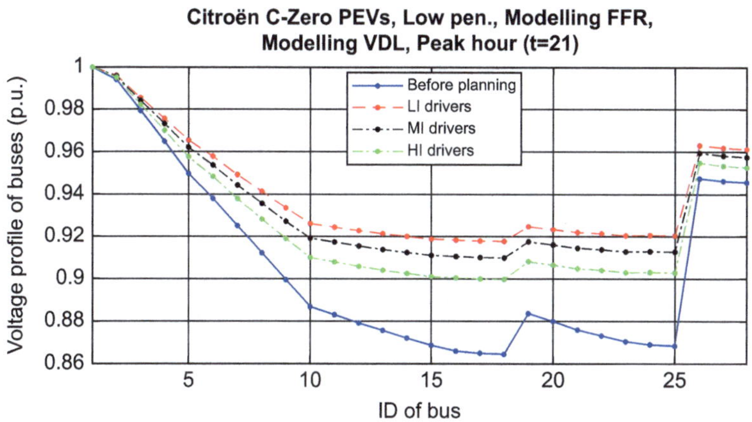

Fig. 6.30 The primary and final voltage profile of system buses (p.u.) at peak hour (21st hour) of typical day for different social classes of drivers considering Citroën C-Zero PEVs and low PEV penetration level and modelling FFR (linear) and VDL

Figure 6.31 exhibits the thermal capacity of branches of feeder (MVA) and the apparent power flowing through them (MVA) at peak hour (21st hour) of typical day before and after optimal allocation and sizing of parking lots for each social class of drivers. As can be seen, the security issues of branches with the IDs of 1, 2, 5, and 6 are removed after optimal placement and sizing of parking lots for any social class of drivers. However, the cooperation of low-income drivers for the V2G service provision further moderates the power flow of branches and put the system in the most secure status.

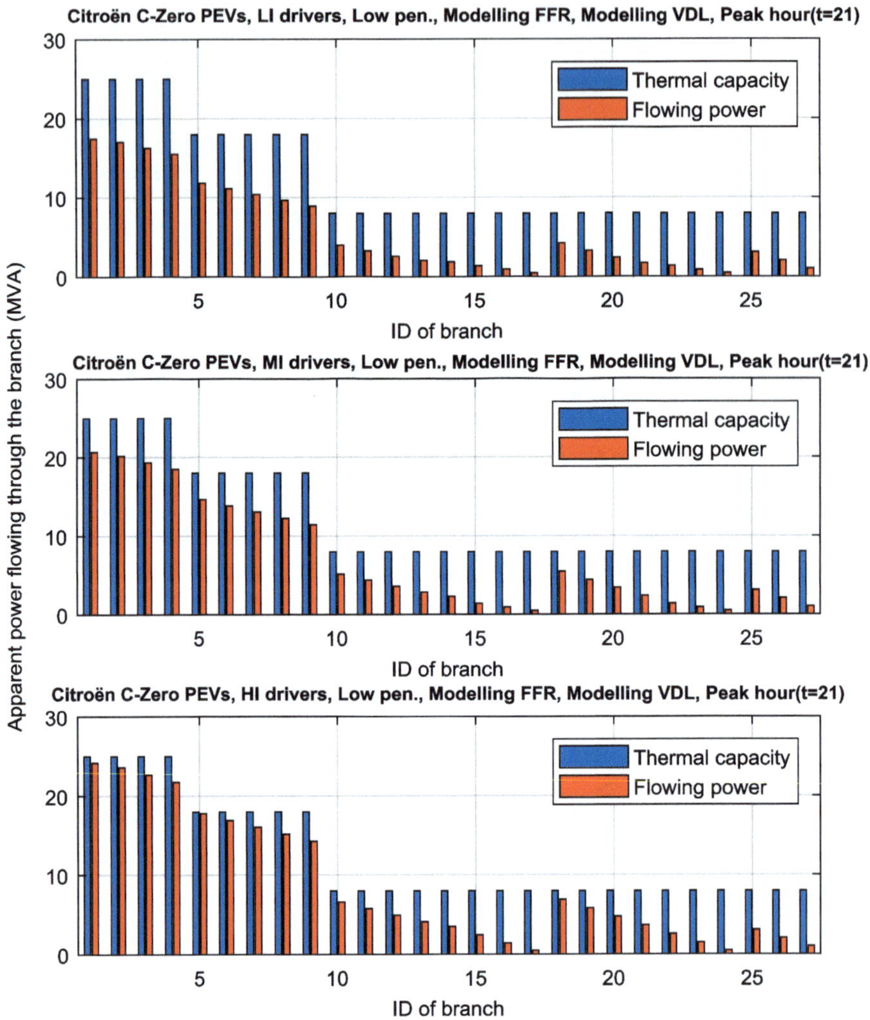

Fig. 6.31 The thermal capacity of branches of feeder (MVA) and apparent power flowing through the branches (MVA) at peak hour (21st hour) of typical day for different social classes of drivers considering Citroën C-Zero PEVs and low PEV penetration level and modelling FFR (linear) and VDL

6.7.4 *Studying the Effects of PEV Penetration Level*

This part of problem simulation is concerned with the fifth-seventh scenarios. In this part, the effects of PEV penetration level (low, moderate, and high) on the problem outputs are investigated considering Citroën C-Zero PEVs and moderate-income drivers, as well as modelling FFR (linear model) and VDL.

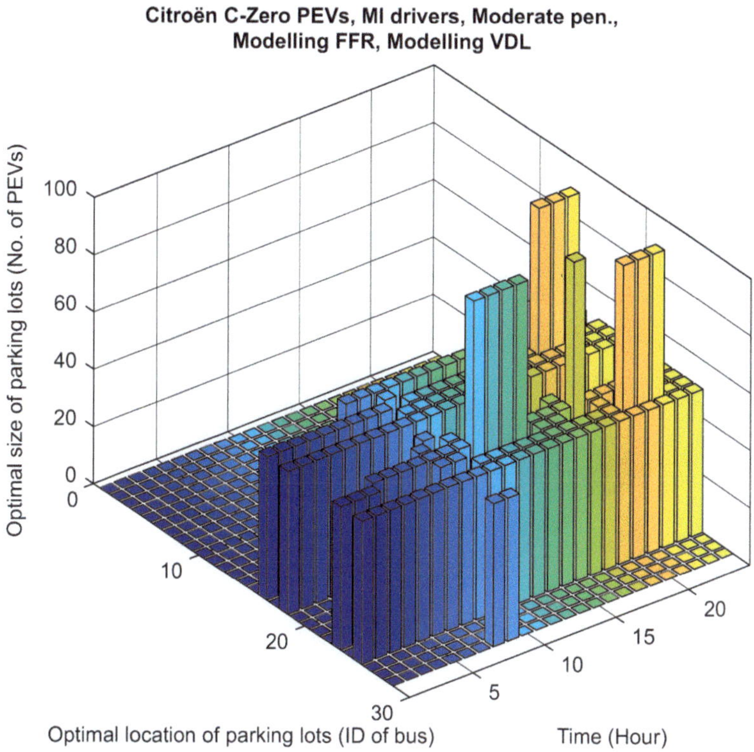

Fig. 6.32 The optimal hourly location and size of parking lots in the typical day for moderate PEV penetration level considering Citroën C-Zero PEVs and moderate-income drivers and modelling FFR (linear) and VDL

Figures 6.32 and 6.33, in addition to Fig. 6.22, show the optimal hourly location and size of parking lots in the typical day for the moderate, high, and low PEV penetration levels, respectively. By comparing them, it is noticed that even the PEV penetration level can affect the planning problem outputs. As can be seen, the more PEV penetration level, the more and larger parking lots are allocated to the electrical distribution grid.

The optimal hourly value of incentive (%) in the typical day for the different PEV penetration levels is illustrated in Fig. 6.34. As can be noticed, lower optimal incentive needs to be considered in the case of high PEV penetration level, compared to other penetration levels.

Figure 6.35 displays the sensitivity analysis for the planning problem cost ($) at peak hour (21st hour) of typical day with respect to the value of incentive (%) for various PEV penetration levels. As is shown, each PEV penetration level is concerned with a single and distinctive optimal point corresponding to the minimum cost and optimal incentive. Furthermore, the high PEV penetration level is desirable, since it results in the lowest minimum cost.

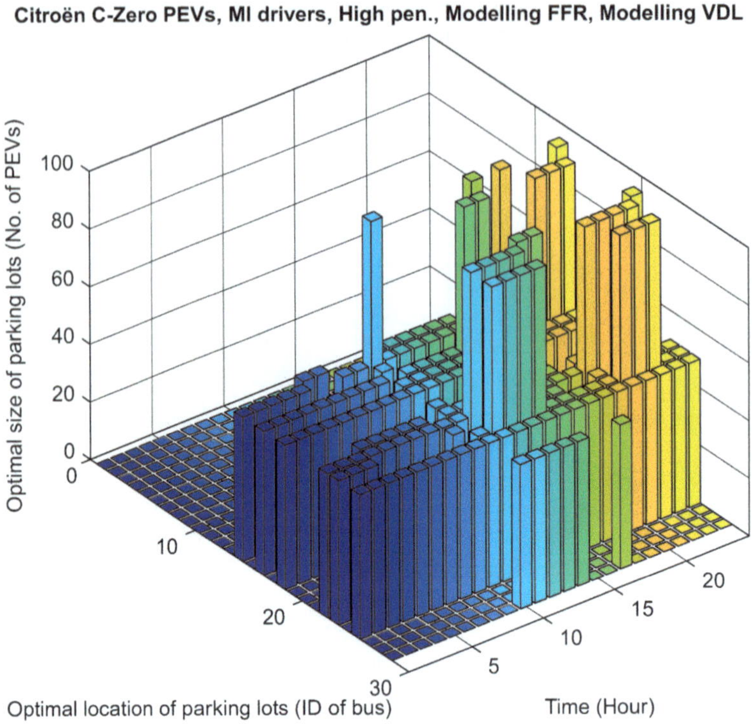

Fig. 6.33 The optimal hourly location and size of parking lots in the typical day for high PEV penetration level considering Citroën C-Zero PEVs and moderate-income drivers and modelling FFR (linear) and VDL

The primary and final energy loss and ENS of system during the planning period (GWh) for various PEV penetration levels are shown in Fig. 6.36. The figure indicates that the high penetration level of PEV causes the lowest energy loss and ENS during the planning time horizon.

The primary and final reliability indices of system in the typical year for different PEV penetration levels are exhibited in Fig. 6.37. As is seen in the charts, the more PEV penetration level, the better value for the reliability indices is achieved. It should be noted that SAIFI, SAIDI, and AENS are the unfavorable indices and ASAI is a desirable index.

Figure 6.38 shows the primary and final total cost and cost terms of problem during the planning period ($) for dissimilar PEV penetration levels. As is observed, the high penetration level of PEV results in the lowest minimum total cost of planning problem, although it imposes the most incentive, investment, and maintenance costs.

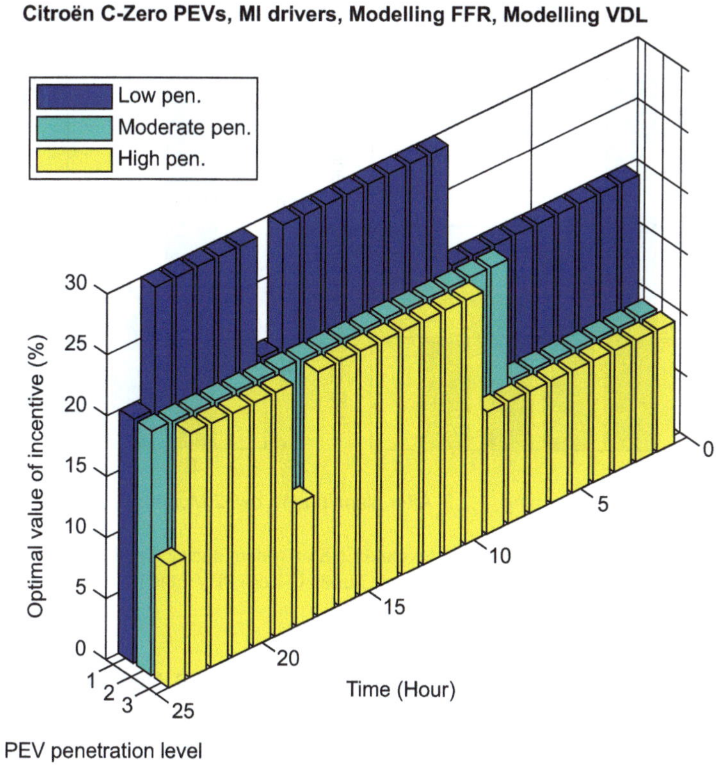

Fig. 6.34 The optimal hourly value of incentive (%) in the typical day for various PEV penetration levels considering Citroën C-Zero PEVs and moderate-income drivers and modelling FFR (linear) and VDL

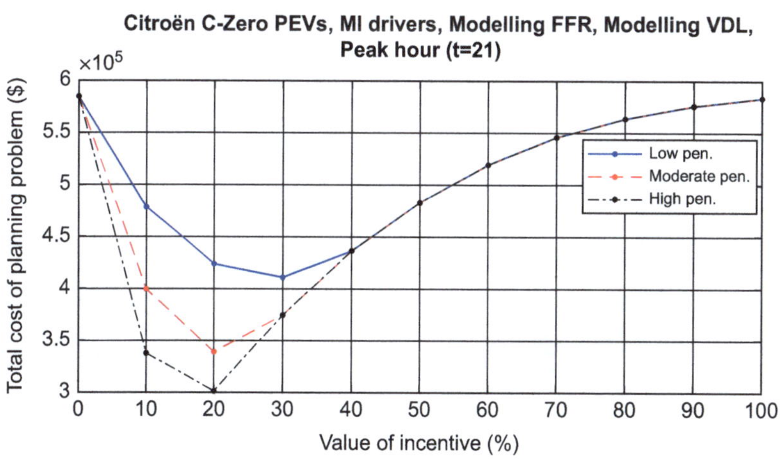

Fig. 6.35 The sensitivity analysis for the planning problem cost ($) at peak hour (21st hour) of typical day with respect to the value of incentive (%) for various PEV penetration levels considering Citroën C-Zero PEVs and moderate-income drivers and modelling FFR (linear) and VDL

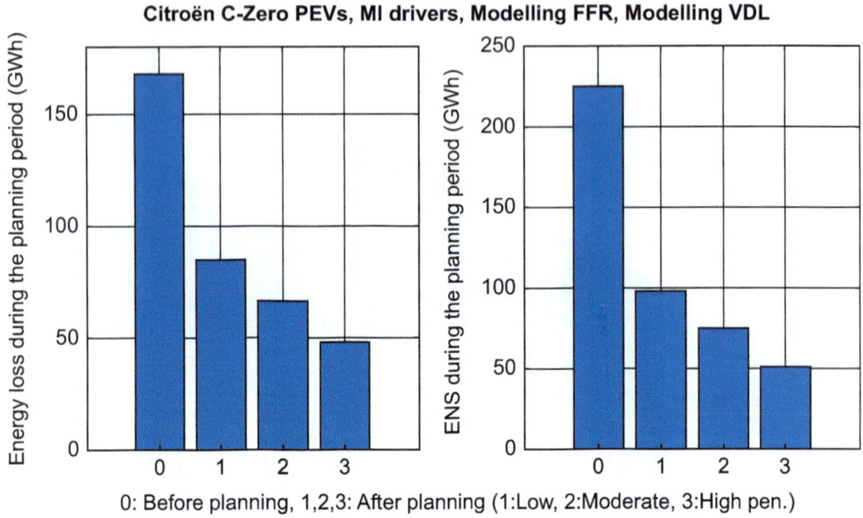

Fig. 6.36 The primary and final energy loss and ENS of system during the planning period (GWh) for various PEV penetration levels considering Citroën C-Zero PEVs and moderate-income drivers and modelling FFR (linear) and VDL

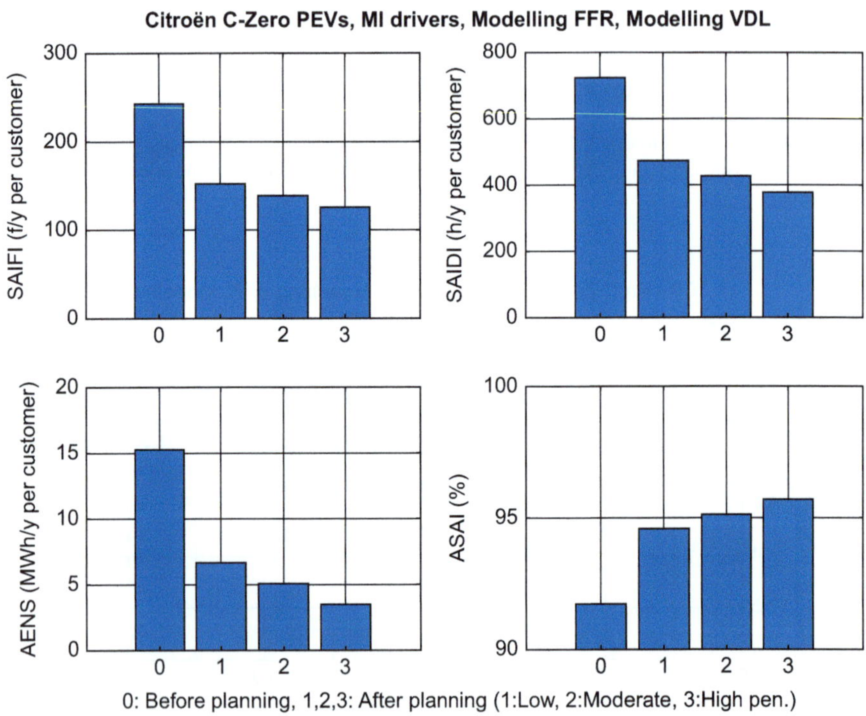

Fig. 6.37 The primary and final reliability indices of system in the typical year for various PEV penetration levels considering Citroën C-Zero PEVs and moderate-income drivers and modelling FFR (linear) and VDL

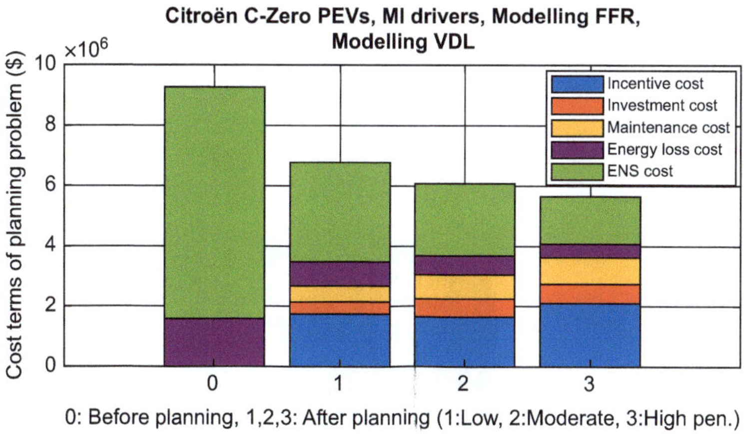

Fig. 6.38 The primary and final total cost and cost terms of problem during the planning period ($) for various PEV penetration levels considering Citroën C-Zero PEVs and moderate-income drivers and modelling FFR (linear) and VDL

6.7.5 Studying the Effects of PEV Type

In this part, the eighth scenario is considered to study the effects of PEV type on the planning problem. In this scenario, Tesla Model S PEV is taken into consideration, and the problem simulation results are compared with the ones of Citroën C-Zero PEV. Herein, moderate-income drivers and low PEV penetration level are considered, and FFR (linear model) and VDL are modelled.

Figure 6.39 shows the optimal hourly location and size of parking lots in the typical day for Tesla Model S PEVs. By comparing this figure with Fig. 6.22, it is noticed that PEV type can affect the optimal location and size of parking lots, and Tesla Model S PEVs need fewer and smaller parking lots. This phenomenon is concerned with the rated power and capacity of PEVs' battery. In other words, to optimally provide the V2G service in the electrical distribution system, the local DISCO needs fewer Tesla Model S PEVs, due to their larger rated power and capacity.

Figure 6.40 illustrates the optimal hourly value of incentive (%) in the typical day for both PEV types. As can be noticed from the charts, the drivers of Tesla Model S PEVs need less optimal incentive to cooperate with the local DISCO for V2G service provision.

The sensitivity analysis for the planning problem cost ($) at peak hour (21st hour) of typical day with respect to the value of incentive (%) for different PEV types is presented in Fig. 6.41. As can be seen, the PEV type is able to affect the optimal point, where Tesla Model S PEVs have a better performance.

The energy loss and ENS of system during the planning period (GWh) and the reliability indices of system in the typical year for both PEV types are illustrated in Fig. 6.42. As can be seen, due to the larger rated power and capacity of battery of a

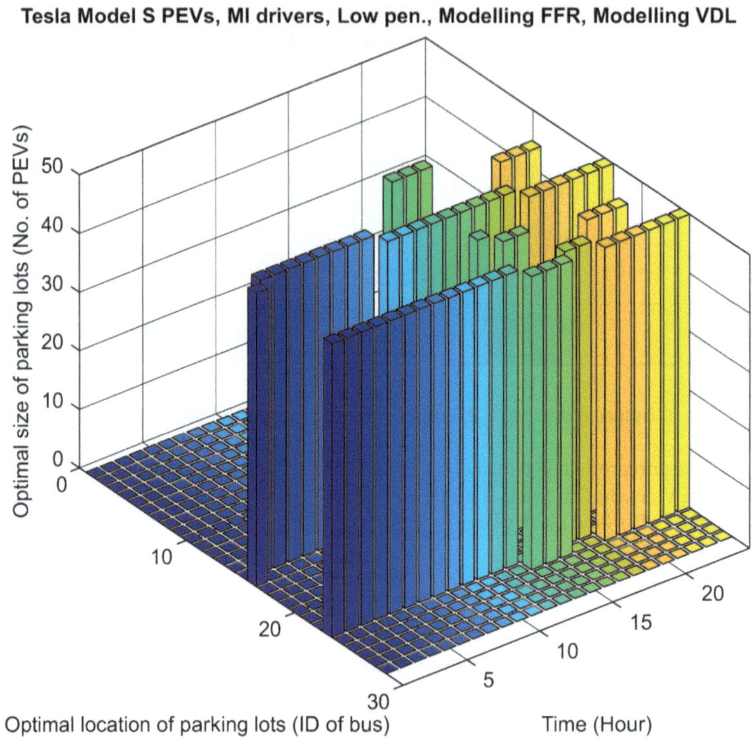

Fig. 6.39 The optimal hourly location and size of parking lots in the typical day for Tesla Model S PEVs considering moderate-income driver and low PEV penetration level and modelling FFR (linear) and VDL

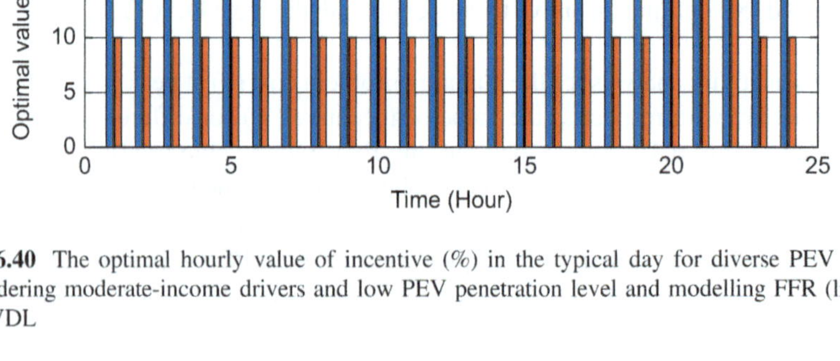

Fig. 6.40 The optimal hourly value of incentive (%) in the typical day for diverse PEV types considering moderate-income drivers and low PEV penetration level and modelling FFR (linear) and VDL

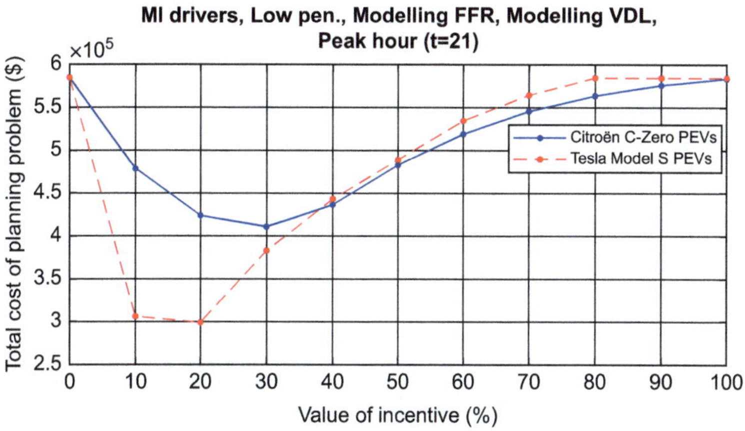

Fig. 6.41 The sensitivity analysis for the planning problem cost ($) at peak hour (21st hour) of typical day with respect to the value of incentive (%) for diverse PEV types considering moderate-income drivers and low PEV penetration level and modelling FFR (linear) and VDL

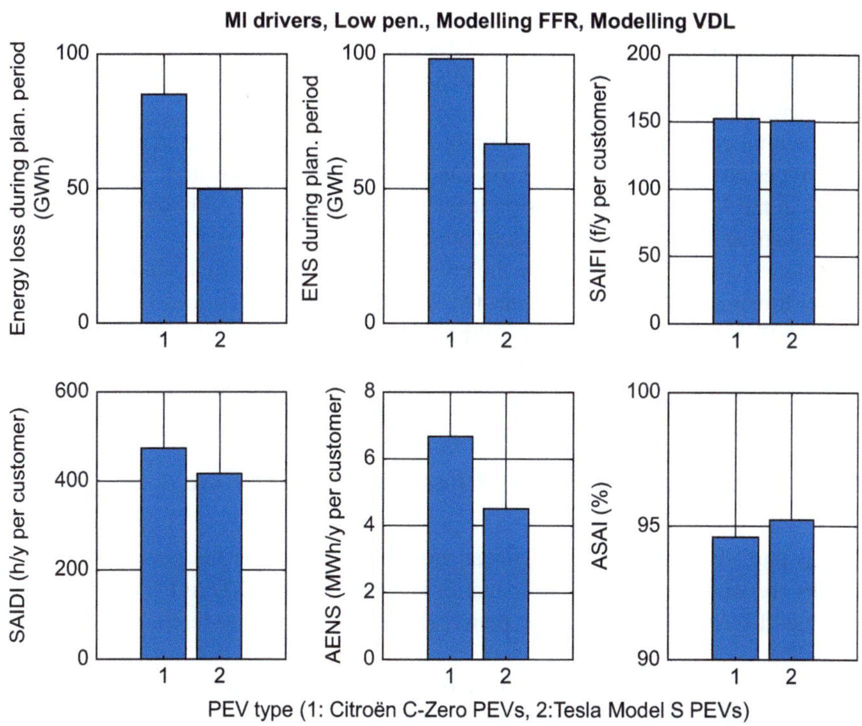

Fig. 6.42 The energy loss and ENS of system during the planning period (GWh) and the reliability indices of system in the typical year for different PEV types considering moderate-income drivers and low PEV penetration level and modelling FFR (linear) and VDL

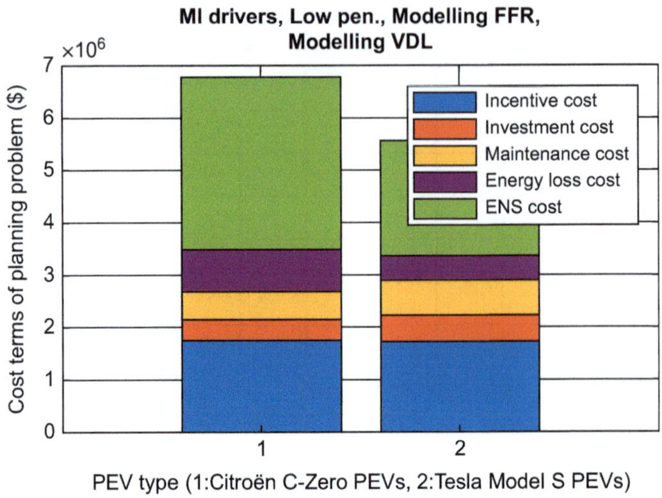

Fig. 6.43 The total cost and cost terms of problem during the planning period ($) for diverse PEV types considering moderate-income drivers and low PEV penetration level and modelling FFR (linear) and VDL

Tesla Model S PEV, the value of all the undesirable parameters, except SAIFI, that is, energy loss, ENS, SAIDI, and AENS, decreases, and the value of desirable parameter (ASAI) increases.

The total cost and cost terms of problem during the planning period ($) for both PEV types are given in Fig. 6.43. As can be seen, the application of Tesla Model S PEVs in the V2G service provision leads to the better result, although the value of investment and maintenance costs increase in this case.

Figures 6.44 and 6.45 illustrate the voltage profile of buses (p.u.) and the apparent power flowing through the branches (MVA) at peak hour (21st hour) of typical day for both PEV types. As can be seen, the system has a better security status in the case of application of Tesla Model S PEVs.

6.7.6 *Studying the Effects of Feeder's Failure Rate*

In this part, the effects of FFR (linear model) modelling on the planning problem results are studied, as the ninth scenario. Herein, Citroën C-Zero PEVs, moderate-income drivers, and low PEV penetration level are considered, and VDL is modelled.

Figures 6.46, 6.47, 6.48, 6.49, 6.50, 6.51, and 6.52 show the value of error in the problem simulation results due to disregarding the real model of FFR of branches (i.e., considering constant model instead of linear model). As is seen, due to the ignorance of real model of FFR (linear model), misleading results are achieved for the optimal hourly location and size of parking lots in the typical day (see Fig. 6.46),

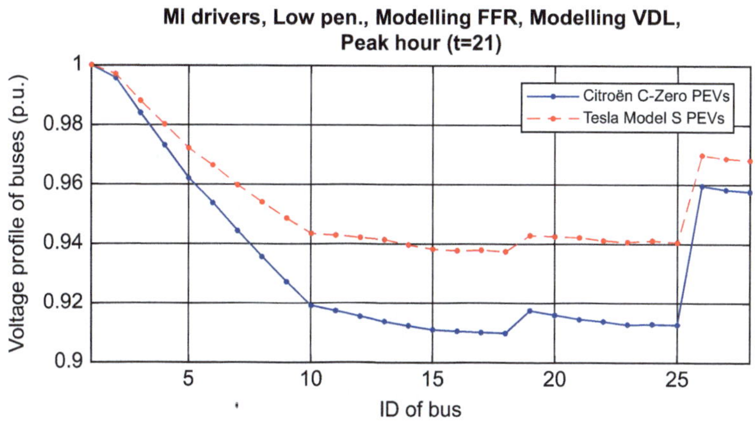

Fig. 6.44 The voltage profile of buses (p.u.) at peak hour (21st hour) of typical day for various PEV types considering moderate-income drivers and low PEV penetration level and modelling FFR (linear) and VDL

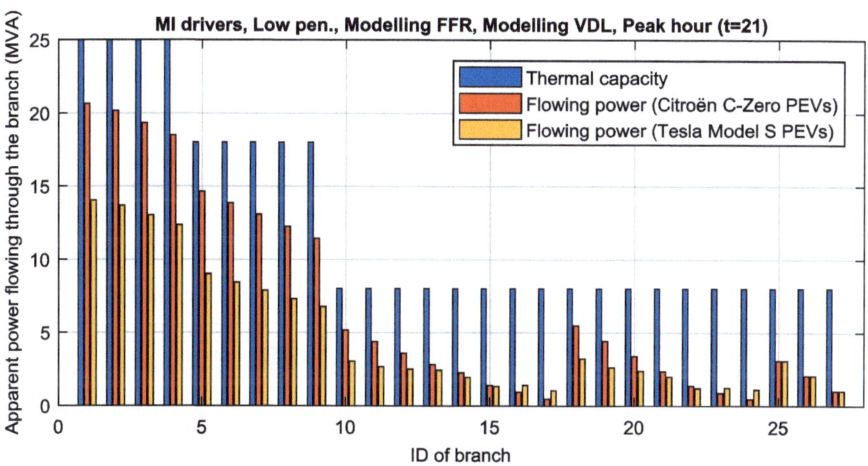

Fig. 6.45 The thermal capacity of branches of feeder (MVA) and apparent power flowing through the branches (MVA) at peak hour (21st hour) of typical day for both PEV types considering moderate-income drivers and low PEV penetration level and modelling FFR (linear) and VDL

as well as for the optimal hourly value of incentive in the typical day (see Fig. 6.47). Herein, the positive and negative values of error show that the results are misleadingly higher and lower than the actual amounts, respectively.

In addition, Figs. 6.48 and 6.49 indicate that, due to the ignorance of real model of FFR (linear model), there is an error in the value of problem simulation outputs that include the energy loss and ENS of system during the planning period (see Fig. 6.48), yearly SAIFI, yearly SAIDI, yearly AENS, and yearly ASAI (see Fig. 6.49).

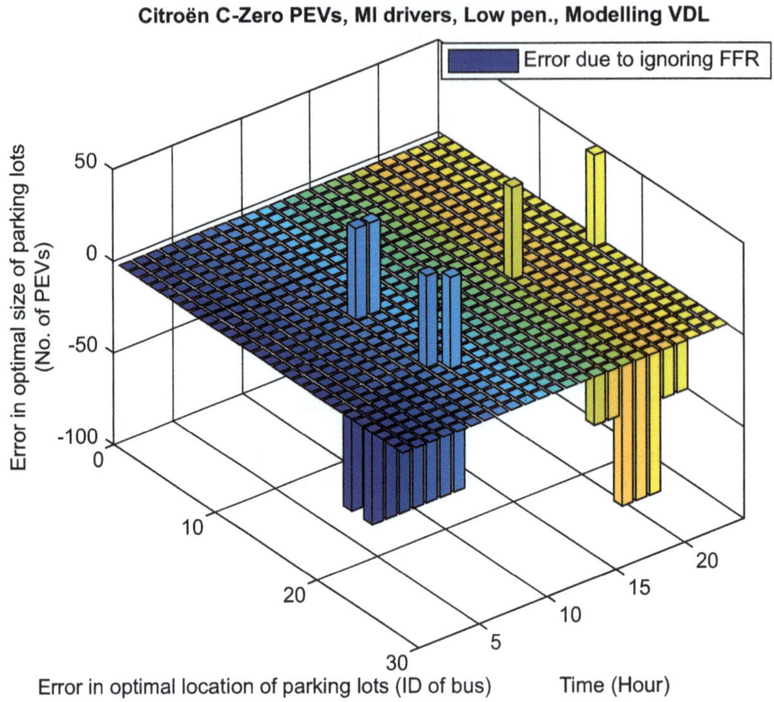

Fig. 6.46 The value of error in the optimal hourly location and size of parking lots in the typical day due to ignoring the real model of FFR (linear), considering Citroën C-Zero PEVs, moderate-income drivers, and low PEV penetration level, as well as modelling VDL

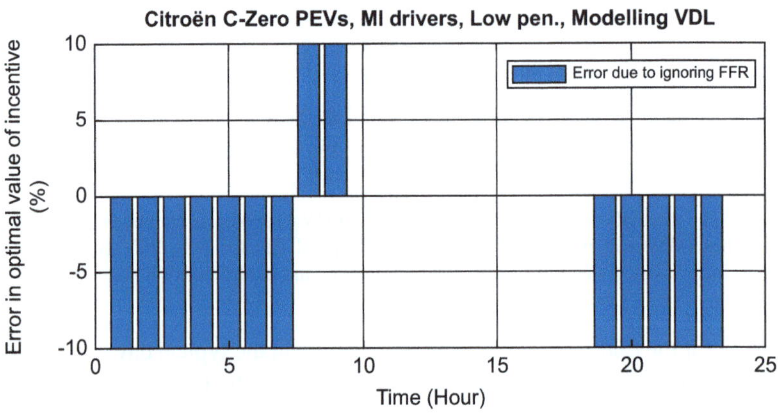

Fig. 6.47 The value of error in the optimal hourly value of incentive (%) in the typical day due to ignoring the real model of FFR (linear), considering Citroën C-Zero PEVs, moderate-income drivers, and low PEV penetration level, as well as modelling VDL

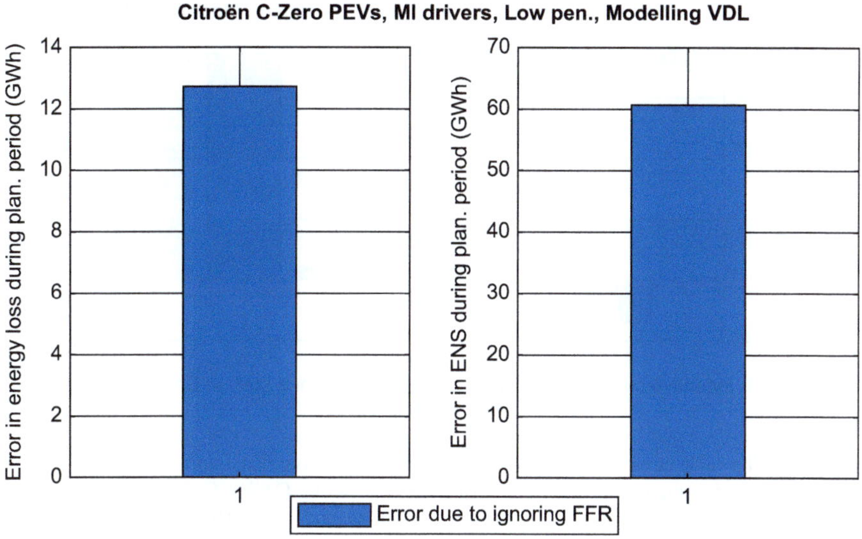

Fig. 6.48 The value of error in the energy loss and ENS of system during the planning period (GWh) due to ignoring the real model of FFR (linear), considering Citroën C-Zero PEVs, moderate-income drivers, and low PEV penetration level, as well as modelling VDL

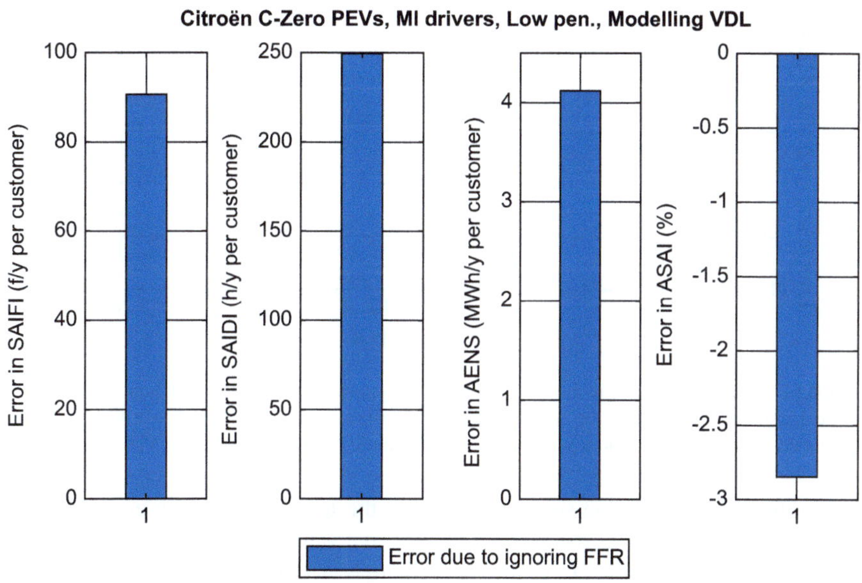

Fig. 6.49 The value of error in the reliability indices of system in the typical year due to ignoring the real model of FFR (linear), considering Citroën C-Zero PEVs, moderate-income drivers, and low PEV penetration level, as well as modelling VDL

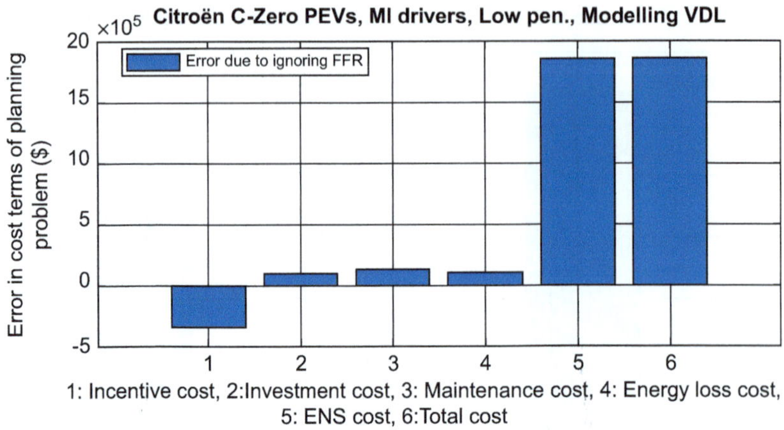

Fig. 6.50 The value of error in the total cost and cost terms of problem during the planning period ($) due to ignoring the real model of FFR (linear), considering Citroën C-Zero PEVs, moderate-income drivers, and low PEV penetration level, as well as modelling VDL

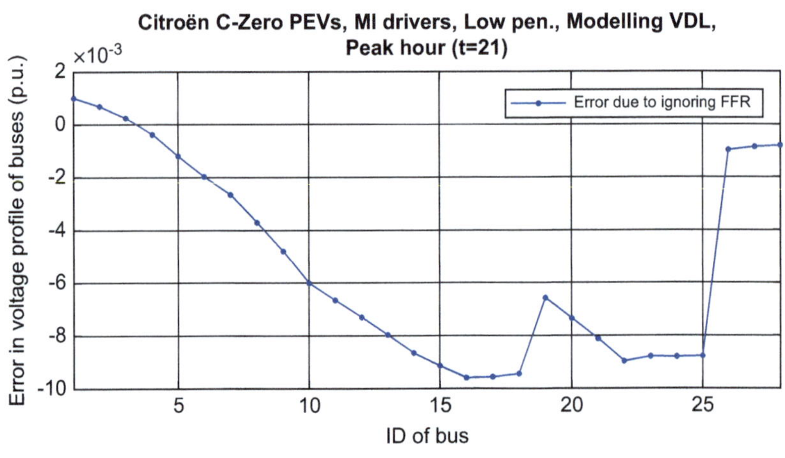

Fig. 6.51 The value of error in the voltage profile of system buses (p.u.) at peak hour (21st hour) of typical day due to ignoring the real model of FFR (linear), considering Citroën C-Zero PEVs, moderate-income drivers, and low PEV penetration level, as well as modelling VDL

Furthermore, as can be seen in Fig. 6.50, the total cost of problem during the planning period increases (because of the positive value of error), and a detriment is imposed to the local DISCO. Additionally, there is an error in the value of cost terms, especially in the ENS cost, since this index has a direct relation with the FFR.

Moreover, the security status of system worsens, since the voltage profile of system at peak hour of typical day (see Fig. 6.51) and the apparent power flowing through the branches at peak hour of typical day (see Fig. 6.52) show the negative

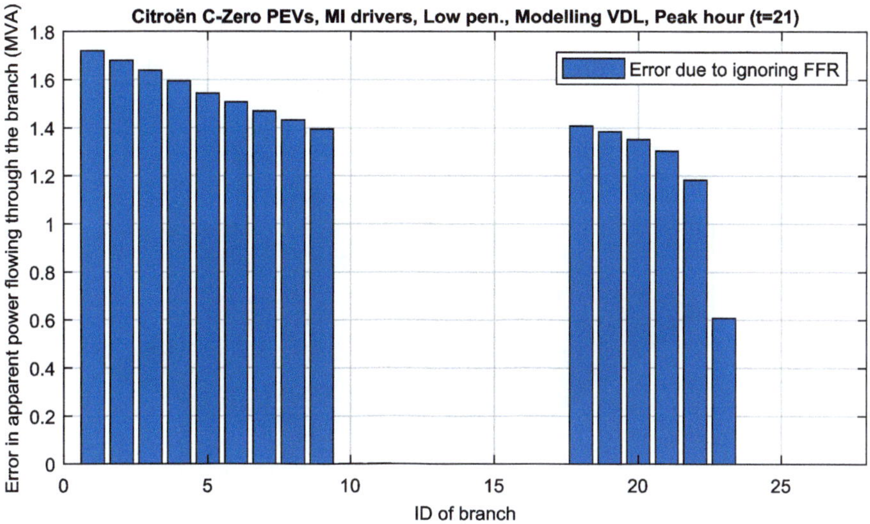

Fig. 6.52 The value of error in the apparent power flowing through the branches of feeder (MVA) at peak hour (21st hour) of typical day due to ignoring the real model of FFR (linear), considering Citroën C-Zero PEVs, moderate-income drivers, and low PEV penetration level, as well as modelling VDL

and positive values of error, respectively. In other words, the voltage profile of buses drops, and the power flow of branches grows, due to the ignorance of real model of FFR of branches (i.e., considering constant model instead of linear model).

6.7.7 Studying the Effects of Voltage-Dependent Load

In this part, the tenth scenario is studied, where the effects of VDL modelling on the problem results are analyzed. Herein, Citroën C-Zero PEVs, moderate-income drivers, and low PEV penetration level are considered, and FFR (linear model) is modelled.

Because of neglecting the VDL modelling, an error is observed in the problem simulation results, presented in Figs. 6.53, 6.54, 6.55, 6.56, and 6.57. As can be seen, nonoptimal and deceptive consequences are obtained for the hourly location and size of parking lots and the hourly value of incentive in the typical day. Moreover, the problem simulation outputs in Fig. 6.55 show that the value of energy loss, cost terms, and total cost during the planning period increases due to the ignorance of VDL modelling. Additionally, a considerable error is observed in the voltage profile of system and the system power flow, as can be seen in Figs. 6.56 and 6.57.

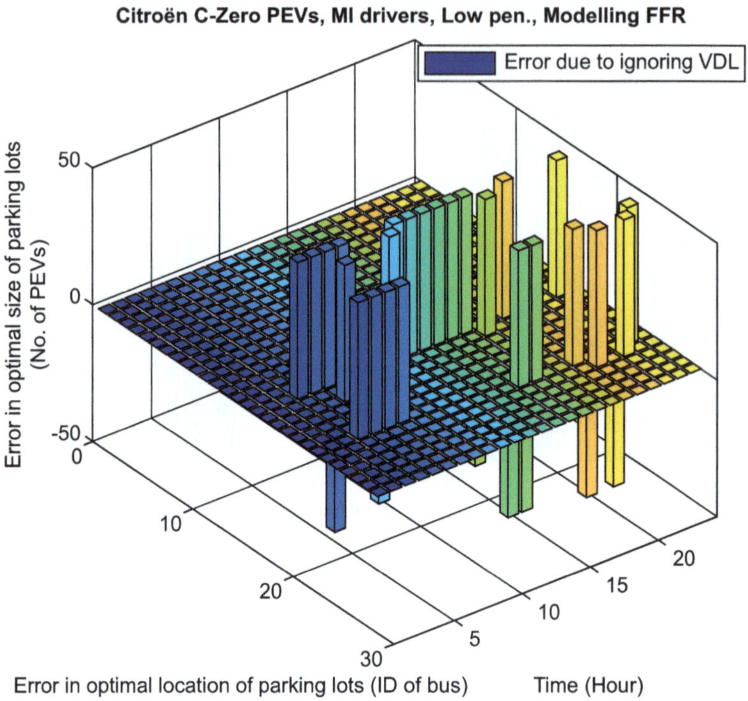

Fig. 6.53 The value of error in the optimal hourly location and size of parking lots in the typical day due to ignoring the VDL modelling, considering Citroën C-Zero PEVs, moderate-income drivers, and low PEV penetration level, as well as modelling FFR (linear)

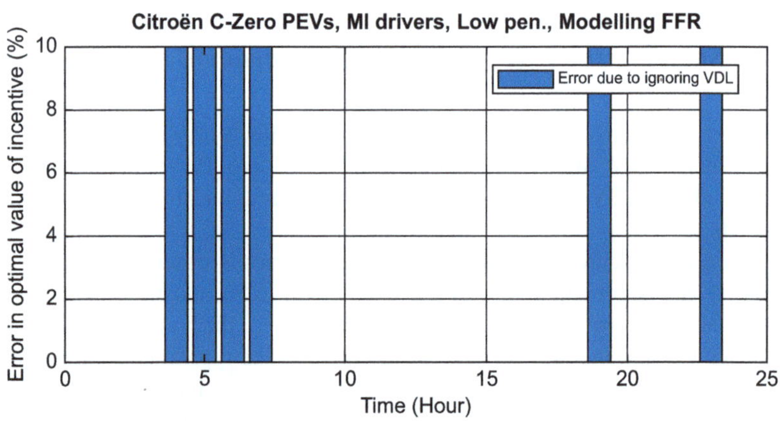

Fig. 6.54 The value of error in the optimal hourly value of incentive (%) in the typical day due to ignoring the VDL modelling, considering Citroën C-Zero PEVs, moderate-income drivers, and low PEV penetration level, as well as modelling FFR (linear)

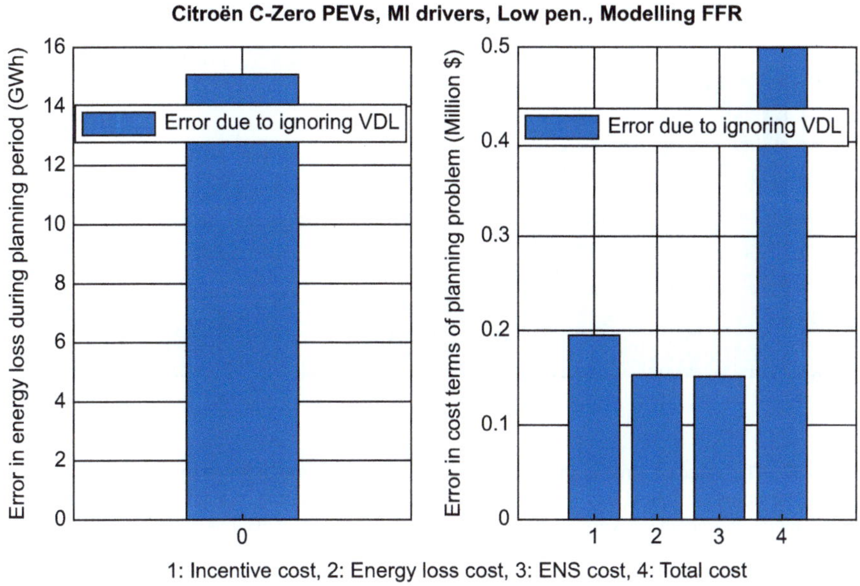

Fig. 6.55 The value of error in the energy loss (MWh), cost terms, and total cost of problem during the planning period ($) due to ignoring the VDL modelling, considering Citroën C-Zero PEVs, moderate-income drivers, and low PEV penetration level, as well as modelling FFR (linear)

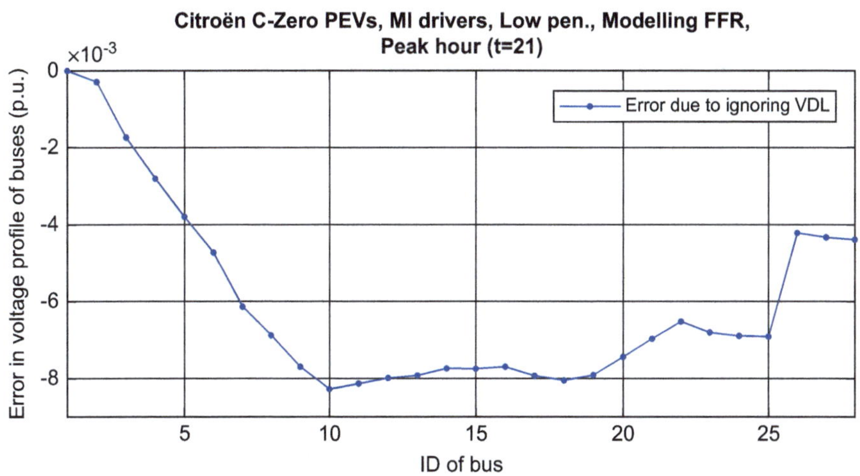

Fig. 6.56 The value of error in the voltage profile of system buses (p.u.) at peak hour (21st hour) of typical day due to ignoring the VDL modelling, considering Citroën C-Zero PEVs, moderate-income drivers, and low PEV penetration level, as well as modelling FFR (linear)

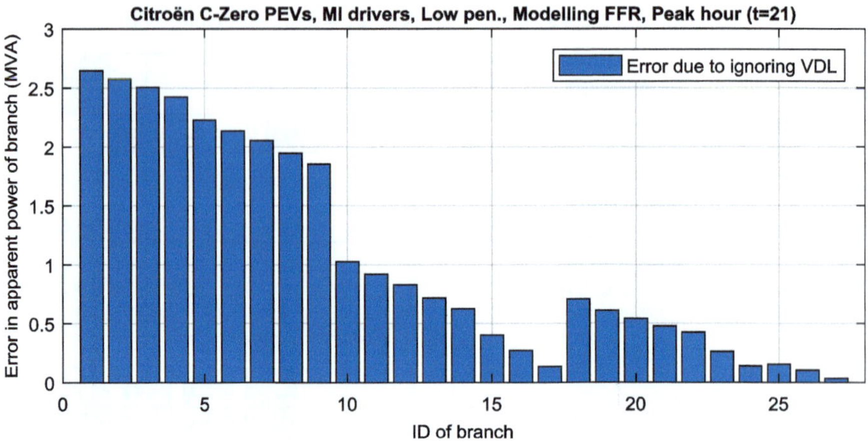

Fig. 6.57 The value of error in the apparent power flowing through the branches (MVA) at peak hour (21st hour) of typical day due to ignoring the VDL modelling, considering Citroën C-Zero PEVs, moderate-income drivers, and low PEV penetration level, as well as modelling FFR (linear)

6.8 Conclusion

The planning problem of parking lot sizing and placement in the electrical distribution system considering the security constraints and modelling the technical, geographical, and social aspects was studied in this chapter. The problem was investigated from the local distribution company's (DISCO) viewpoint with the aim of minimum total cost of problem during the planning period considering the economic factors, the real driving routes of vehicles in San Francisco, the social class of drivers, the plug-in electric vehicle (PEV) penetration level, and the PEV type, as well as modelling the feeder's failure rate (FFR) and voltage-dependent load (VDL).

In this chapter, several scenarios were investigated to study and analyze the effects of input parameters on the output parameters. The input parameters included the social classes of drivers (low-income, moderate-income, and high-income), the PEV penetration levels (low, moderate, and high), the PEV types (Citroën C-Zero and Tesla Model S), the FFR models (constant and linear), and the VDL models (residential, commercial, and industrial). In addition, the output parameters included the optimal hourly location and size of parking lots, the optimal hourly value of incentive, the minimum total cost of planning problem, the energy loss and energy not supplied (ENS) of system during the planning period, the reliability indices of system in the typical year, the voltage profile of buses, and the power flow of feeder's branches. The other models of FFR, that is, power, exponential, and logarithmic models, were not studied in this chapter; thus they are suggested as the future study.

The optimal value of incentive was demonstrated, and the optimal size and location of parking lots need to be updated during the day. In addition, it was proven that all the abovementioned objective parameters can affect the optimal value of output parameters.

Regarding the effects of social class of drivers on the planning problem outputs, it was noticed that the least and the most incentives are needed to motivate the low-income and high-income drivers to participate in the vehicle-to-grid (V2G) service provision, respectively. Additionally, the largest and the smallest parking lots are allocated to the electrical distribution network for the low-income and high-income drivers, respectively. Moreover, their participation in the V2G service provision resulted in the best and the worst values of objective function, energy loss, reliability indices, voltage profile of buses, and power flow of branches, respectively.

Concerning the effects of PEV penetration level on the problem outputs, it was proven that for higher PEV penetration level, lower optimal incentives need to be offered to the drivers. Furthermore, in the case of high level of PEV penetration, the best results are achieved.

With respect to the effects of PEV type on the output parameters, fewer and smaller parking lots are needed to install in the system for the PEVs that have larger battery (Tesla Model S). In addition, their drivers need less incentive to cooperate with the local DISCO to provide V2G service. Moreover, it was shown that the application of this type of PEV in the V2G service provision results in the better results and the system has a superior security status.

Regarding the ignorance of FFR and VDL modelling, it was seen that nonoptimal values are assigned for the incentive, as well as for the location and the size of parking lots, and misleading results are achieved for the problem outputs. Moreover, because of neglecting the abovementioned modelling, the total cost of planning problem increases, some detriments are imposed to the local DISCO, and the security status of system is negatively affected.

Nomenclature

α	Distance of a PEV from the parking lot (meter)
β	Value of incentive (percent) offered to the drivers to motivate them to provide V2G service at suggested parking lots and proposed intervals
ξ	Responsiveness probability of a driver (percent)
φ	Phase angle of admittance of a branch in polar coordinate system
δ	Phase angle of voltage of a bus in polar coordinate system
ρ_1, ρ_2	Exponents of active and reactive power demands of an electricity consumer in the VDL model
λ_0	Failure rate of a branch before parking lot allocation (f/y)
λ_f	Failure rate of a branch after fully compensating the active power demand of two buses of branch (f/y)
λ	Failure rate of a branch (f/y)
λ^{CON}	Failure rate of a branch considering the constant model of FFR (f/y)
λ^{LIN}	Failure rate of a branch considering the linear model of FFR (f/y)
λ^{POW}	Failure rate of a branch considering the power model of FFR (f/y)
λ^{EXP}	Failure rate of a branch considering the exponential model of FFR (f/y)

(continued)

λ^{LOG}	Failure rate of a branch considering the logarithmic model of FFR (f/y)
λ^{Model}	Failure rate of a branch considering an FFR model (f/y)
U^{Model}	Annual outage time of a bus (h/y)
μ	A certain percentage of investment cost (%) defined as the yearly maintenance cost of a parking lot
π^{E}	Electricity price ($/MWh)
π^{CD}	Customer's damage due to power outage ($/MWh)
θ^{M}	Mutation probability of a gene in GA (%)
$AENS$	Average energy not supplied index (MWh/year per customer)
$ASAI$	Average system availability index (%)
b, B, N^{B}	Index of branch, set of branches of feeder, and total number of branches of feeder
c, C, N^{C}	Index of customer, set of customers, number of customers connected to the bus
C^{PL}	Building cost of a parking lot and equipping it for one PEV ($)
$Cost^{INV}$	Investment cost to build the parking lots and equip them with the charging/discharging stations ($)
d, D	Index of day and set of days of a year
DOD^{PEV}	Depth of discharge defined for a PEV (%)
e, $PEVs$, N^{PEVs}	Index of PEV, set of PEVs, and total number of PEVs in the area
E^{LOSS}	Energy loss of feeder's branches (MWh)
Eff^{PEV}	Efficiency of a PEV's battery for V2G power provision (%)
ENS	Energy not supplied index (MWh/year)
f_{ch}	Fitness of a chromosome in GA
i, j, J, N^{J}	Index of bus, index of bus, set of buses, and total number of buses of system
$\|I\|$	Magnitude of current flowing through a branch (ampere)
IFR	Inflation rate (%)
ITR	Interest rate (%)
LNS	Load not supplied (MW)
LNS^{FL}	Load not supplied during a fault locating process (MW)
LNS^{FR}	Load not supplied during a fault repairing process (MW)
$min\|V\|$, $max\|V\|$	Lower and upper limits of voltage magnitude of a bus (p.u.)
$Model$	Set of FFR models including constant (CON), linear (LIN), power (POW), exponential (EXP), and logarithmic (LOG) models
$\|MVA\|$	Magnitude of apparent power flowing through the branch (MVA)
n^{PL}	Step to change the size of a parking lot
n^{PL-MAX}	Maximum step to change the size of a parking lot
N^{GA}	Number of chromosomes of population in GA
OF	Objective function of planning problem ($)
P_0, P	Operating and rated active power demands (MW)
$\widetilde{P^{V2G}}$	Expected V2G power of a PEV (MW)
P^{PEV}	Rated power of a PEV's battery (kW)
$PWV(Cost^{M})$	Present worth value of maintenance cost of parking lots and their equipment during the planning period ($)

(continued)

PWV ($Cost^{INC}$)	Present worth value of incentives paid to the drivers to motivate them to provide the V2G service during the planning period ($)
PWV ($Cost^{LOSS}$)	Present worth value of energy loss cost of feeder's branches during the planning time horizon ($)
PWV ($Cost^{ENS}$)	Present worth value of ENS cost due to the faults and failures occurred in the feeder's branches during the planning period ($)
Q_0, Q	Operating and rated reactive power demands (MVAr)
R	Resistance of a branch (ohm)
Rand[0, 1)	A random number in the range of [0,1) in GA
Res, *Com*, *Ind*	Indices of residential, commercial, and industrial electricity consumers
$Real\{I_0\}$	Real component of current flowing through the branch before parking lot allocation (ampere)
$Real\{I\}$	Real component of current flowing through a branch after fully compensating the active power demand of two buses of branch (ampere)
SAIDI	System average interruption duration index (hour/year per customer)
SAIFI	System average interruption frequency index (failure/year per customer)
Selection	Selection status of a chromosome in the probabilistic fitness-based selection process in GA
$Size^{PL}$	Size or capacity of a parking lot (number of PEVs)
$Size^{PL-MIN}$	Minimum size of a parking lot (number of PEVs)
SC, *LI*, *MI*, *HI*	Indices of a driver's social class, low-income social class, moderate-income social class, and high-income social class
SOC^{PEV}	State of charge of a PEV (%)
t, T	Index of time and set of hours of a day
TC	Thermal capacity of a branch (MVA)
T^{FL}	Failure locating duration (hour)
T^{FR}	Failure repairing duration (hour)
V_0, V	Operating and rated voltage profile of a bus (p.u.)
$\|V_0\|$, $\|V\|$	Magnitude of operating and rated voltage profile of a bus (p.u.)
y, Y, N^Y	Index of year, set of years of planning period, and length of planning time horizon (year)
$\|Y\|$	Magnitude of admittance of line between two buses

Abbreviations

AENS	Average energy not supplied
ASAI	Average system availability index
C.B.	Circuit breaker
CCS	Combined Charging System
Com	Commercial
CON	Constant
DICSO	Distribution company
DOD	Depth of discharge
ENS	Energy not supplied

(continued)

EXP	Exponential
F1, F2	Feeder 1, feeder 2
FFR	Feeder's failure rate
GA	Genetic algorithm
IFR	Inflation rate
Ind	Industrial
ITR	Interest rate
LI, MI, HI	Low-income, moderate-income, high-income
LIN	Linear
LOG	Logarithmic
PEV	Plug-in electric vehicle
POW	Power
R.C.	Recloser
Res	Residential
SAIDI	System average interruption duration index
SAIFI	System average interruption frequency index
SOC	State of charge
V2G	Vehicle-to-grid
VDL	Voltage-dependent load

References

1. N. Lutsey, Global climate change mitigation potential from a transition to electric vehicles, International Council on Clean Transportation, 2015. [Online]. Available: http://theicct.org/global-ev-2050-ghg-mitigation-potential. Accessed on Jan 2019
2. [Online]. Available: https://www.theicct.org/sites/default/files/publications/ICCT_EV-promotion-US-cities_20150729.pdf. Accessed on Jan 2019
3. [Online]. Available: https://www.ucsusa.org/clean-vehicles/electric-vehicles/car-charging-time-type-cost#.XGLy1zNKjIU. Accessed on Jan 2019
4. [Online]. Available: https://afdc.energy.gov/fuels/electricity_locations.html#/find/nearest?fuel=ELEC&ev_connectors=NEMA1450&ev_connectors=NEMA515&ev_connectors=NEMA520&ev_connectors=J1772&ev_levels=dc_fast. Accessed on Jan 2019
5. A. Gagarin, P. Corcoran, Multiple domination models for placement of electric vehicle charging stations in road networks. Comput. Oper. Res. **96**, 69–79 (2018)
6. M. Rahmani-Andebili, Optimal power factor for optimally located and sized solar parking lots applying quantum annealing. IET Gener. Transm. Distrib. **10**, 2538–2547 (2016)
7. S.M. Mohammadi-Hosseininejad, A. Fereidunian, A. Shahsavari, H. Lesani, A healer reinforcement approach to self-healing in smart grid by PHEVs parking lot allocation. IEEE Trans. Ind. Info. **12**(6), 2020–2030 (2016)
8. S.M. Mohammadi-Hosseininejad, A. Fereidunian, H. Lesani, Reliability improvement considering plug-in hybrid electric vehicles parking lots ancillary services: A stochastic multi-criteria approach. IET Gener. Transm. Distrib. **12**(4), 824–833 (2018)
9. M. Rahmani-Andebili, G.K. Venayagamoorthy, SmartPark placement and operation for improving system reliability and market participation. Electr. Pow. Syst. Res. (Elsevier) **123** (6), 21–30 (2015)
10. M. Rahmani-Andebili, H. Shen, M. Fotuhi Firuzabad, Planning and operation of parking lots considering system, traffic, and drivers behavioral model. IEEE Trans. Syst. Man Cybern. Syst. (2018). https://doi.org/10.1109/TSMC.2018.2824122

11. M. Rahmani-Andebili, H. Shen, Traffic and grid-based parking lot allocation for PEVs considering driver behavioral model, in *International Conference on Computing, Networking and Communications (ICNC)*, Silicon Valley, USA, 26–29 Jan 2017
12. M. Rahmani-Andebili, M. Fotuhi Firuzabad, An adaptive approach for PEVs charging management and reconfiguration of electrical distribution system penetrated by renewables. IEEE Transactions on Industrial Informatics **14**(5), 2001 (2018)
13. [Online]. Available: https://www.usgs.gov/faqs/how-much-distance-does-a-degree-minute-and-second-cover-your-maps. Accessed on Jan 2019
14. [Online]. Available: https://crawdad.org/all-byname.html. Accessed on Jan 2019
15. M. Rahmani-Andebili, Economic and network-driven distributed generation placement modeling Feeder's failure rate and Customer's load type. IEEE Trans. Ind. Electron. **63**(3), 1598–1606 (2016)
16. M. Rahmani-Andebili, Simultaneous placement of DG and capacitor in distribution network. Electr. Pow. Syst. Res. (Elsevier) **131**, 1–10 (2016)
17. M. Rahmani-Andebili, Reliability and economic-driven switchable capacitor placement in distribution network. IET Gener. Transm. Distrib. **9**(13), 1572–1579 (2015)
18. M. Rahmani-Andebili, Effect of load models on optimal capacitor allocation in distribution network, in *IEEE Electric Power Distribution Networks Conference (EPDC)*, Tehran, Iran, pp. 1–4, 2–3 May 2012
19. IEEE Task Force on Load Representation for Dynamic Performance, Bibliography on load models for power flow and dynamic performance simulation. IEEE Trans. Power Syst. **10**, 523–538 (1995)
20. R.E. Brown, *Electric Power Distribution Reliability* (Marcel Dekker Inc., New York/Basel, 2002)
21. P.L. Lewin, J.E. Theed, A.E. Davies, S.T. Larsen, Method for rating power cables buried in surface troughs. IEE Proc., Gener. Transm. Distrib. **146**, 360–364 (1999)
22. [Online]. Available: http://www.elektriauto.ee/eng.y. Accessed on Jan 2019

Chapter 7
Estimating the State of Charge of Plug-In Electric Vehicle Fleet Applying Monte Carlo Markov Chain

7.1 Introduction

Plug-in electric vehicles (PEV) can mitigate or even remove the energy security and environmental issues of transportation sector, since they can be powered by the renewable energy sources which are much cleaner and cheaper than the fossil fuels. Even if the PEVs are charged by the electricity generated by the fossil fuels, they will still produce less pollution than their conventional counterparts, that is, internal combustion engine vehicles [1].

Figure 7.1 presents the 2018 monthly and total sales of electric vehicles in the USA for the major automakers [2]. As can be seen, Tesla, Toyota, Honda, Chevrolet, Nissan, BMW, Ford, Chrysler, Mitsubishi, and Kia are the top electric vehicle sellers in the USA. In addition, Tesla Model 3, Toyota Prius Prime, and Tesla Model X are the most popular and best-selling electric vehicles in the USA in 2018.

As can be seen, the electric vehicle sales in the USA has increased from 199,826 to 361,307 in 2018. Moreover, Fig. 7.1 indicates that the worldwide electric vehicle sales in 2018 was about 1,713,957. Figure 7.2 illustrates the annual global electric vehicle sales predicted by 2030 [3]. As is seen, the total number of electric vehicles was about 1 million in 2017, while it is predicted to hit 24.4 million by 2030 [3]. Furthermore, it is forecasted that China and the USA will have the largest markets of electric vehicles in the future.

Power system operators, power market players, and PEV aggregators are some of the agents that are interested in their own customers' behavior modeling. PEVs' drivers are the new electricity customers that will consume a considerable portion of electricity in the future. In this regard, the SOC of PEV fleet, as one of the most important factors that can affect the modelling of drivers' behavior in terms of charging time, needs to be estimated by the agents.

In the previous studies [4–6], the SOC of PEV fleet has been considered as the given data or directly calculated using some parameters. In this chapter, Monte Carlo Markov Chain (MCMC) is applied to estimate the best-fit line and confidence bands

M. Rahmani-Andebili, *Planning and Operation of Plug-In Electric Vehicles*,
https://doi.org/10.1007/978-3-030-18022-5_7

2018 U.S. EV SALES	JAN	FEB	MAR	APR	MAY	JUN	JUL	AUG	SEP	OCT	NOV	DEC	TOTAL
Tesla Model 3* 🔋	1875	2485	3820	3750	6000	5902	14250	17800	22250	17750	18650	25250	139,782
Toyota Prius Prime	1496	2050	2922	2626	2924	2237	1984	2071	2213	2001	2312	2759	27,595
Tesla Model X* 🔋	700	975	2825	1025	1450	2550	1325	2750	3975	1225	3200	4100	26,100
Tesla Model S* 🔋	800	1125	3375	1250	1520	2750	1200	2625	3750	1350	2750	3250	25,745
Honda Clarity PHEV*	604	911	1131	1129	1639	1495	1542	1462	1997	2025	1897	2770	18,602
Chevrolet Volt*	713	983	1782	1325	1675	1336	1475	1825	2129	1475	2530	1058	18,306
Chevrolet Bolt EV* 🔋	1177	1424	1774	1275	1125	1083	1175	1225	1549	1975	2825	1412	18,019
Nissan LEAF 🔋	150	895	1500	1171	1576	1367	1149	1315	1563	1234	1128	1667	14,715
BMW 530e*	224	413	689	518	729	942	536	749	756	733	1012	1363	8,664
Ford Fusion Energi	640	794	782	742	740	604	522	396	480	453	1131	790	8,074
Chrysler Pacifica Hybrid**	375	450	480	425	650	710	450	654	637	623	895	713	7,062
BMW i3 (BEV + REx) 🔋	382	623	992	503	424	580	464	418	461	424	490	356	6,117
BMWX5 xDrive 40e*	261	596	627	563	499	321	431	264	225	224	213	210	4,434
Mitsubishi Outlander PHEV	300	323	373	273	297	390	350	366	378	309	376	431	4,166
Kia Niro PHEV*	155	246	227	120	218	281	225	346	313	323	619	316	3,389
BMW 330e*	101	142	202	166	150	138	106	192	195	229	373	606	2,600
Audi A3 Sportback e-tron*	145	199	214	189	267	238	220	240	230	210	180	265	2,597
Volvo XC60 PHEV*	109	155	167	141	214	226	185	210	215	180	225	240	2,267
Fiat 500e** 🔋	210	235	285	215	250	225	220	75	94	100	148	193	2,250
Porsche Panamera E-Hybrid*	1	2	49	336	275	168	195	200	210	170	200	230	2,036
Mercedes C350e*	29	172	208	158	166	176	165	170	82	75	80	240	1,721
Hyundai IONIQ PHEV*	22	178	218	180	217	143	180	43	11	128	136	134	1,590
Mini Countryman SE PHEV*	127	100	74	106	163	211	210	128	140	117	74	114	1,564
Volvo XC90 T8 PHEV*	99	106	93	90	126	133	115	125	120	100	130	150	1,387
Volkswagen e-Golf 🔋	178	198	164	128	76	32	18	32	14	62	230	222	1,354
smart ED 🔋	84	90	103	80	110	126	103	108	98	95	100	122	1,219
Kia Soul EV* 🔋	115	163	157	152	133	57	130	33	18	61	61	54	1,134
Porsche Cayenne S-E*	113	121	197	265	59	12	15	45	60	25	35	75	1,022
Mercedes GLE 550e*	44	70	181	93	83	75	85	90	42	28	35	140	966
Kia Optima PHEV*	86	103	156	142	98	83	90	39	17	51	79	21	965
Honda Clarity BEV* 🔋	153	74	48	39	34	86	102	75	108	106	37	86	948
BMW i8	32	39	47	57	64	45	72	67	55	64	133	97	772
Ford C-Max Energi	234	142	105	57	18	6	4	4	12	0	0	0	582
Mercedes GLC 350e*		5	57	59	64	66	60	65	27	20	24	120	567
Ford Focus Electric 🔋	70	73	137	83	88	50	46	7	4	0	1	1	560
Hyundai Sonata PHEV*	52	54	78	38	67	62	60	20	15	5	5	4	460
Volvo S90 T8 PHEV*	27	29	52	29	30	35	30	40	45	35	40	45	437
Jaguar I-Pace* 🔋										5	165	223	393
Hyundai IONIQ EV* 🔋	49	3	60	7	32	47	35	21	12	21	34	24	345
BMW 740e*	18	23	31	60	17	16	40	18	25	45	18	28	339
Cadillac CT6 PHEV*	6	24	17	42	30	18	26	23	11	12	13	9	231
Mercedes B250e 🔋	40	49	33	7	3	0	0	1	0	1	0	1	135
Mercedes S550e*	13	3	11	9	7	7	8	10	8	5	4	11	96
2018 U.S. Sales Totals	12,009	16,845	26,443	19,623	24,307	25,029	29,598	36,347	44,544	34,074	42,588	49,900	361,307
2017 U.S. Sales Totals	11,004	12,375	18,542	13,367	16,596	17,046	15,540	16,514	21,242	14,315	17,178	26,107	199,826
2018 Worldwide Sales*	82,000	81,000	141,000	128,450	159,346	157,933	144,975	172,400	200,500	208,800	237,553		1,713,957

Fig. 7.1 The 2018 monthly and total sales for the major automakers [2]. ∗ Estimated sales numbers. ∗∗ Estimated based on the state/rebate data and other reports. BEV models are designated with the battery icon

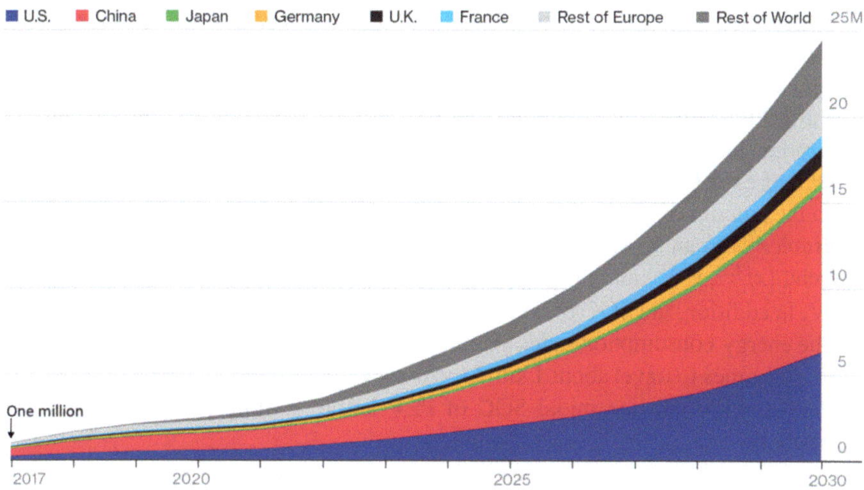

Fig. 7.2 The annual global electric vehicle sales forecasted by 2030 [3]

of hourly SOC of PEV fleet in the typical day. MCMC is a stochastic process to analyze the dataset and determine the probability distribution function of model parameters by repeatedly applying the dataset. The primary dataset applied in this chapter includes the real momentary values of longitude and latitude of driving routes of cars in San Francisco that have been recorded in every 4-minute interval of the day. Herein, due to the lack of dataset about the mobility of PEVs, it is assumed that the PEVs' drivers commute on the same routes.

Then, the PEVs' position dataset is converted to the distance travelled by them. After that, the hourly SOC of PEV fleet is determined applying the technical characteristics of PEVs that include the initial SOC, the energy consumption index, and the capacity of PEVs' batteries. Finally, the best-fit line and confidence bands of hourly SOC of PEV fleet in the given day are estimated using MCMC. In addition, in this chapter, the effects of problem parameters on the MCMC simulation results are studied.

The remainder of paper is organized as follows. In Sect. 7.2, the primary dataset used in the study is presented. In Sect. 7.3, the method of analysis is presented and described. Section 7.4 is concerned with the problem simulation and results analysis, and finally Section 7.5 concludes the paper.

7.2 Dataset

The dataset used in this study is related to the cars' mobility dataset recorded in San Francisco [7]. This dataset includes the position data of cars in longitude (degree) and latitude (degree) recorded in each 4-minute interval of the period of 8:00–24:00. In this study, in order to extract the travelling distance of a car from its longitude-latitude-based position data, one longitude degree travel and one latitude degree travel need to be multiplied by 87.870 km and 111.45 km, respectively, according to reference [8]. In addition, in this chapter, the cars are assumed to be PEVs that will not affect the generality of problem.

By having the position of PEVs in each time-step of day, the state of charge (SOC) of each PEV can be estimated, since the SOC of a PEV has a direct relation with the amount of distance that it travels in a day. The distance that a PEV travels ($Distance^{PEV}$) can be calculated using the Pythagorean formula, presented in (7.1). Herein, $\left(x_t^{PEV}, y_t^{PEV}\right)$ and $\left(x_{t-1}^{PEV}, y_{t-1}^{PEV}\right)$ indicate the current and previous positions of PEV. In addition, the SOC of a PEV can be determined using (7.2). Herein, kWh_{km}, as the energy consumption index of a PEV, is the amount of energy (kWh) that the PEV consumes to travel about 1 km. Moreover, C^{PEV} and SOC_0^{PEV} are the capacity of battery of PEV and the initial SOC of PEV before starting its travel in the day, respectively.

$$Distance_t^{PEV} = \sqrt{\left(x_t^{PEV} - x_{t-1}^{PEV}\right)^2 + \left(y_t^{PEV} - y_{t-1}^{PEV}\right)^2} \tag{7.1}$$

$$SOC_\tau^{PEV} = SOC_0^{PEV} - \frac{kWh_{km}}{C^{PEV}} \times \sum_{t=1}^{\tau} Distance_t^{PEV} \tag{7.2}$$

Figure 7.3 illustrates the daily route of a PEV in San Francisco based on its real position data in longitude (degree) and latitude (degree) recorded in every 4-minute

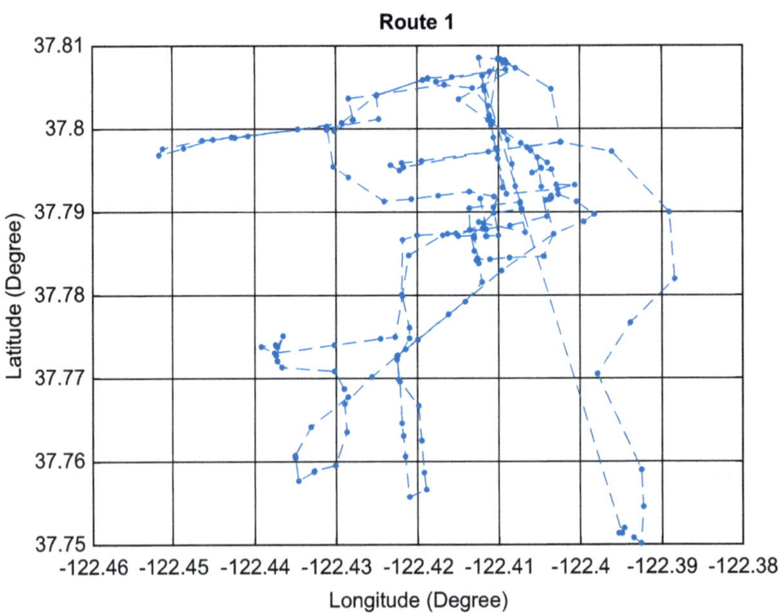

Fig. 7.3 The route of PEV 1 in San Francisco based on its real position data in longitude (degree) and latitude (degree) recorded in each 4-minute interval of period of 8:00–24:00 [7]

interval of the period of 8:00–24:00 [7]. As can be noticed, the vehicle moves with different speeds in each time step, and it is stationary in some periods. The daily routes of 10 PEVs (indicated by routes 1–10) in San Francisco based on their real position data in longitude (degree) and latitude (degree) recorded in every 4-minute interval during the period of 8:00–24:00 are shown in Figs. 7.4, 7.5, and 7.6 [7]. As is seen, each PEV has a unique daily route; thus each PEV travels a different distance in each time step. Therefore, each PEV will have a distinct daily SOC pattern. Figures 7.7, 7.8, and 7.9 show the distances travelled by PEVs 1–10 in every 4-minute interval that have been calculated using Eq. (7.1). Moreover, the hourly distances travelled by the PEVs are illustrated in Figs. 7.10, 7.11, and 7.12.

To determine the hourly SOC of each PEV, the type of PEV needs to be identified. In this chapter, in the base case study, Nissan Leaf is assumed as the PEV type, where its technical specifications are presented in Table 7.1 [9]. As is seen, a Nissan Leaf PEV has a capacity of 24 kWh and it consumes about 0.21 kWh to travel 1 km. Table 7.1 includes the technical specifications of other PEV types that will be studied in the second part of problem simulation [10]. These PEVs are illustrated in Fig. 7.13 [9, 10].

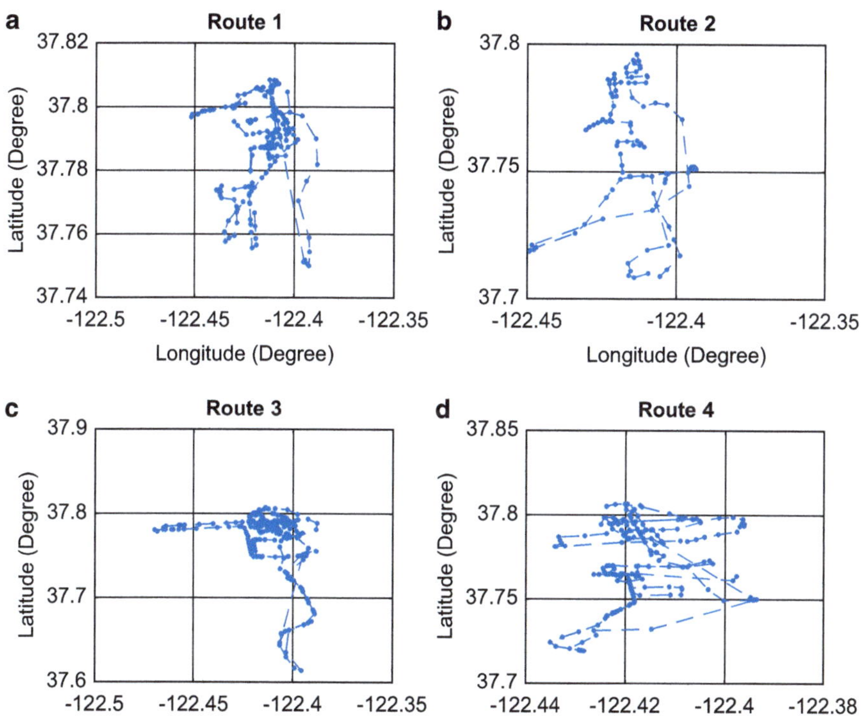

Fig. 7.4 The routes of PEVs (routes 1–4) in San Francisco based on their real position data in longitude (degree) and latitude (degree) recorded in each 4-minute interval of period of 8:00–24:00 [7]

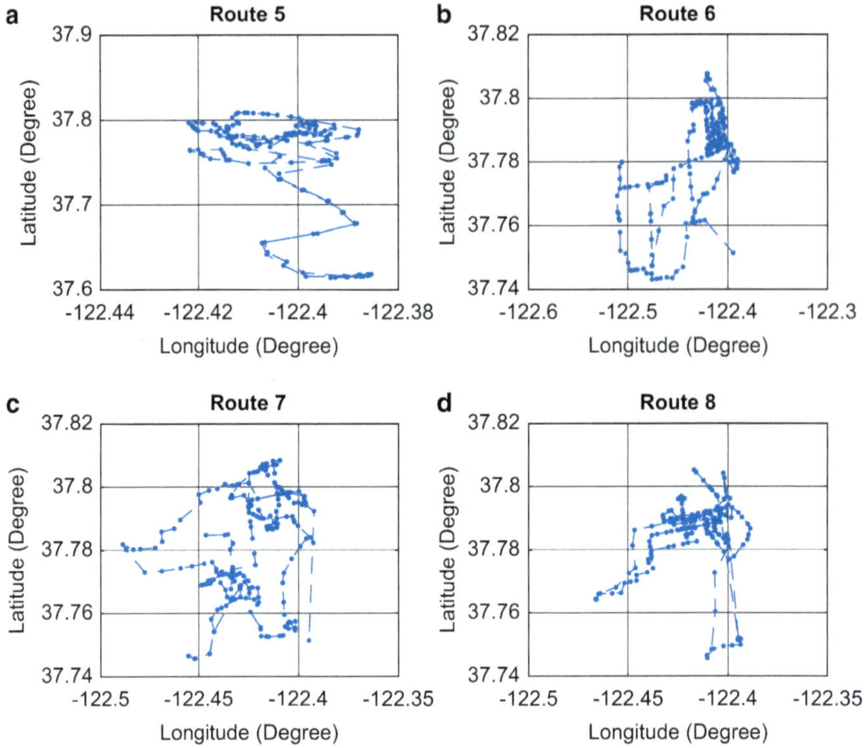

Fig. 7.5 The routes of PEVs (routes 5–8) in San Francisco based on their real position data in longitude (degree) and latitude (degree) recorded in each 4-minute interval of period of 8:00–24:00 [7]

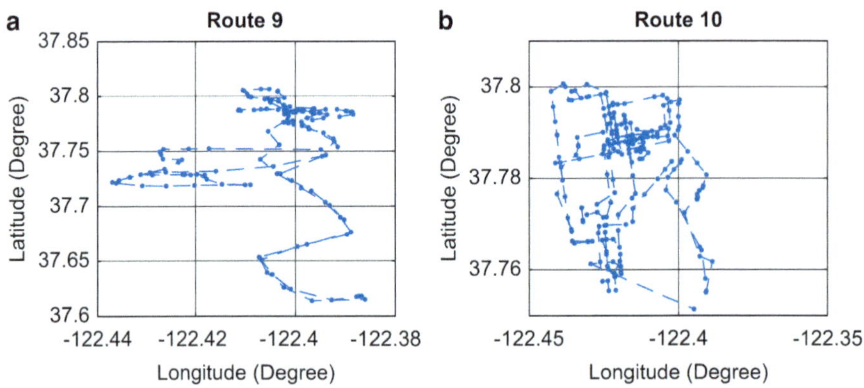

Fig. 7.6 The routes of PEVs (routes 9–10) in San Francisco based on their real position data in longitude (degree) and latitude (degree) recorded in each 4-minute interval of period of 8:00–24:00 [7]

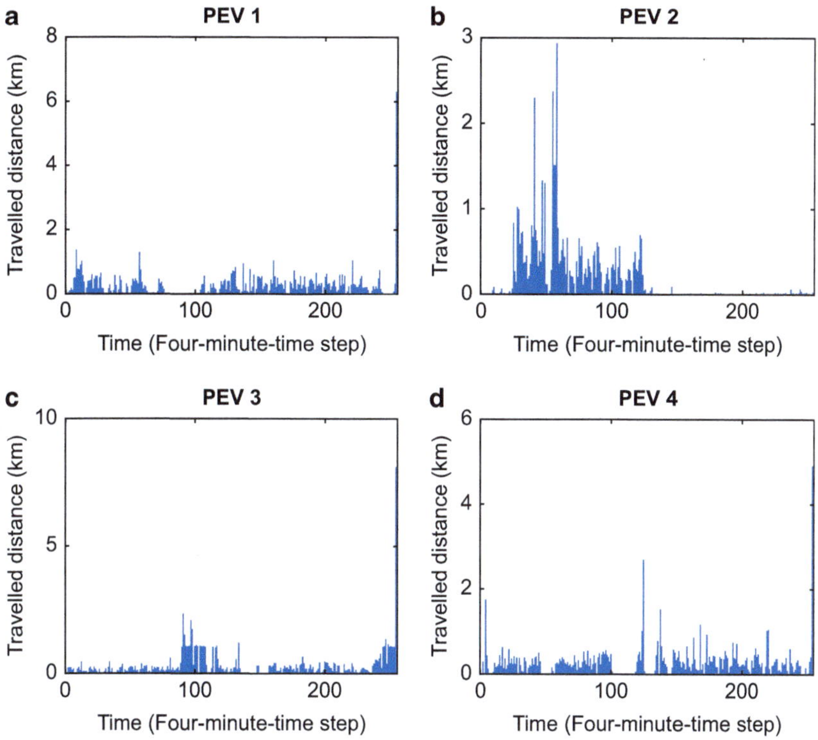

Fig. 7.7 The distances travelled by the PEVs 1–4 calculated in each 4-minute interval of period of 8:00–24:00

Herein, the initial SOC of each PEV is randomly determined in the range of 0.9–1.0 p.u. The hourly SOC of each PEV (p.u.), in the period of 8:00–24:00, is calculated using Eq. (7.2) and illustrated in Figs. 7.14, 7.15, 7.16, and 7.17. Figure 7.18 shows the mean and standard deviation of SOC of PEV fleet (PEVs 1–10) estimated at each hour of period of 8:00–24:00. These data will be used as the input data of MCMC to estimate the best-fit line, mean, standard deviation, and confidence bands of SOC of PEV fleet (p.u.).

7.3 Method of Analysis

MCMC applies a model with some adjustable model parameters ($\theta = \{\theta_1, \ldots, \theta_m\}$) and a proposal distribution function to evaluate the likelihood value ($\mathcal{L}(\theta)$) of set of model parameters to determine the probability distribution function of each model parameter (posterior distribution function). Herein, m is the total number of model parameters. The most common proposal distribution function is a uniform

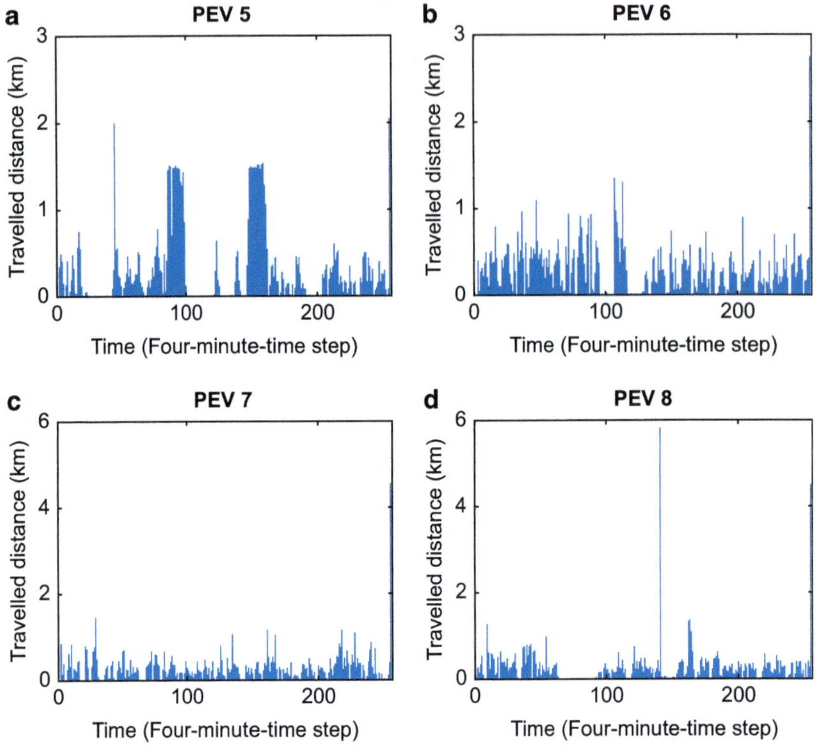

Fig. 7.8 The distances travelled by the PEVs 5–8 calculated in each 4-minute interval of period of 8:00–24:00

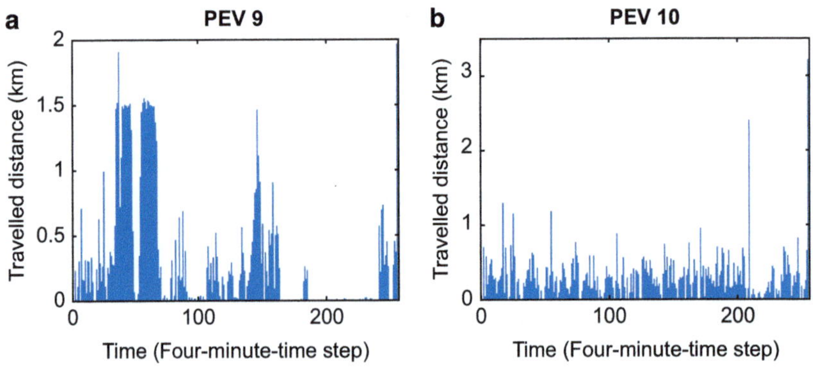

Fig. 7.9 The distances travelled by the PEVs 9–10 calculated in each four-minute interval of period of 8:00–24:00

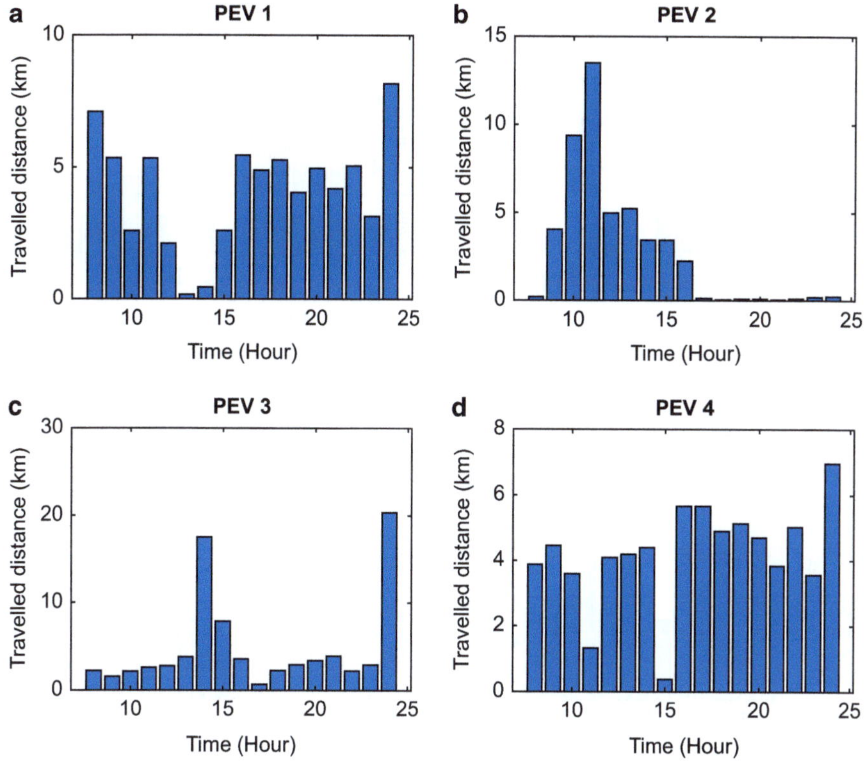

Fig. 7.10 The hourly distances travelled by the PEVs 1–4 calculated for the period of 8:00–24:00

distribution function with a fixed width defined around the current value of model parameter. More information about Markov Chain, MCMC, and stochastic process can be found in [11, 12].

Metropolis-Hastings algorithm (MHA) [13, 14] was devised prior to the mass production of computers. It is one of the most common algorithms that can be applied in MCMC. Herein, the application of MHA in the MCMC is presented and described considering a model with m parameters and a uniform distribution function.

In MHA, first, the initial value of set of model parameters ($\theta^1 = \{\theta_1^1, \ldots, \theta_m^1\}$) is randomly quantified. Next, the likelihood value of current set of model parameters $\left(\mathcal{L}\left(\theta_1^n, \ldots, \theta_m^n\right)\right)$ is determined using Eq. (7.3), where $\theta^n = \{\theta_1^n, \ldots, \theta_m^n\}$, n is the index of iteration of MCMC. In (7.3), i, N, x, y, σ^2, $Y^n\left(x, \theta_1^n, \ldots, \theta_m^n\right)$ indicate the index of dataset, the size of dataset (total number of sample data), the independent variable of dataset, the dependent variable of dataset, the variance of dependent variable of dataset, and the current value of dependent variable of model, respectively. The likelihood value of current set of model parameters can be simplified, as

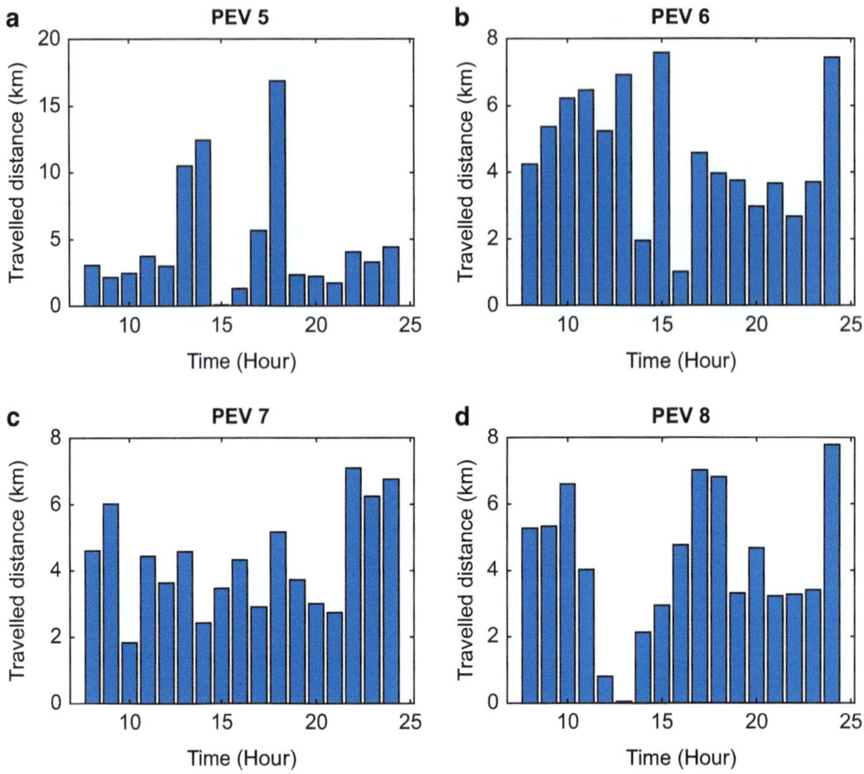

Fig. 7.11 The hourly distances travelled by the PEVs 5–8 calculated for the period of 8:00–24:00

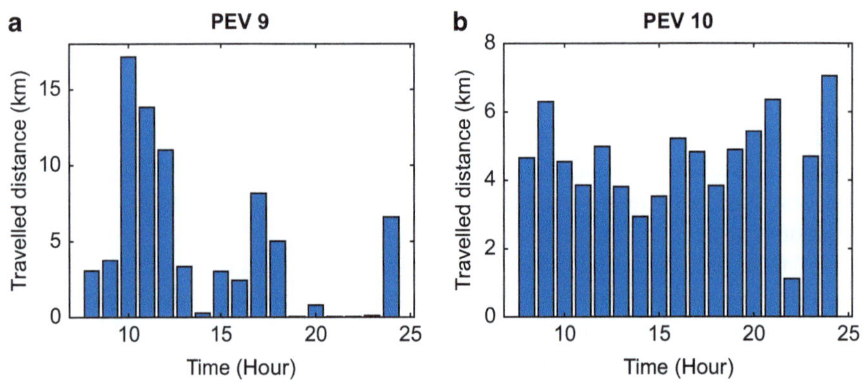

Fig. 7.12 The hourly distances travelled by the PEVs 9–10 calculated for the period of 8:00–24:00

is shown in (7.4). Herein, the Chi-squared variable ($\mathcal{X}^2$) is defined as in Eq. (7.5); thus the likelihood is simplified as in Eq. (7.6). More information about Chi-squared variable can be found in [15–17].

Table 7.1 The technical specifications of some PEV types [9, 10]

Type of PEV	Nissan Leaf [9]	Nissan Leaf S [10]	Mia Electric C3 [10]	Tesla Model S [10]
Energy consumption index (kWh/km)	0.21	0.15	0.1	0.236
Battery capacity (kWh)	24	24	8	60
Charging voltage (V)	230	230	230	230

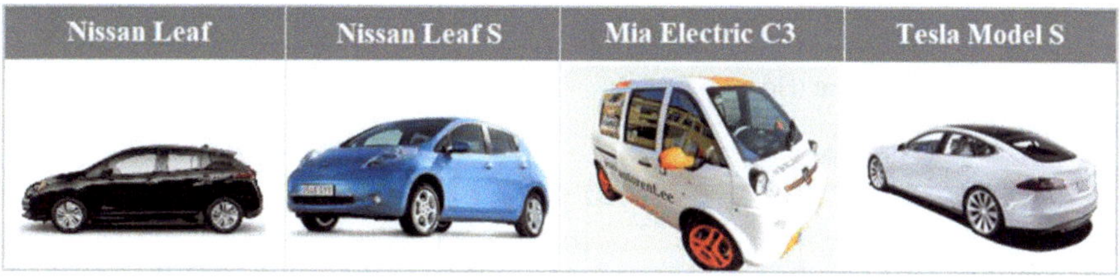

Fig. 7.13 Different PEV types under study [10]

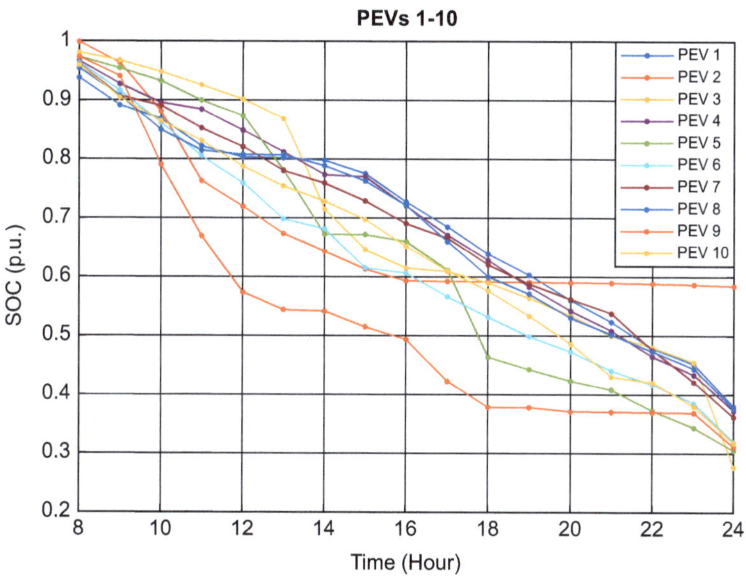

Fig. 7.14 The SOC (p.u.) of PEVs 1–10 estimated at each hour of period of 8:00–24:00

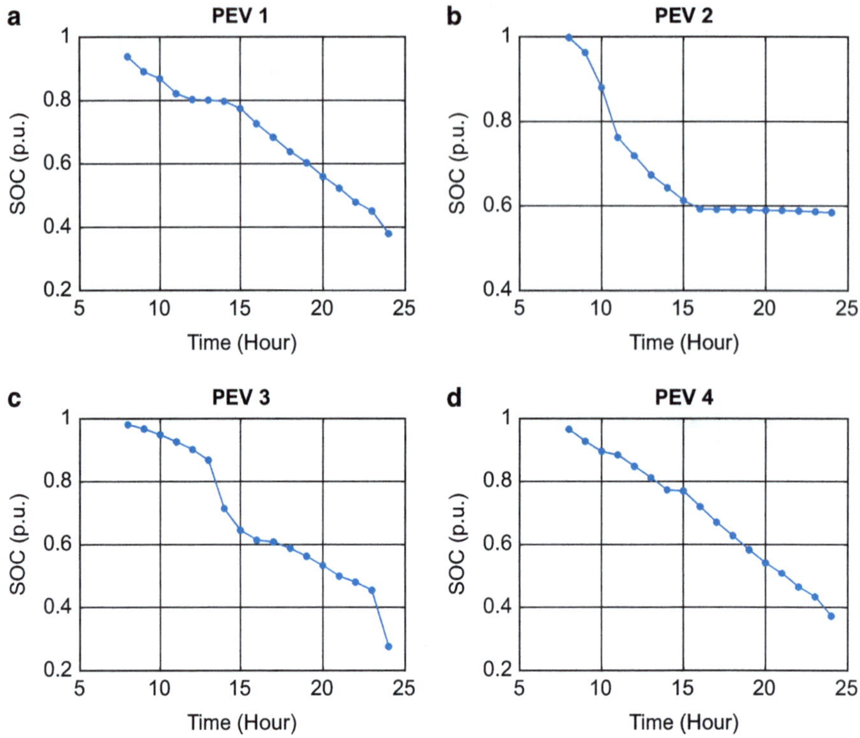

Fig. 7.15 The SOC (p.u.) of PEVs 1–4 estimated at each hour of period of 8:00–24:00

$$\mathcal{L}\left(\theta_1^n, \ldots, \theta_m^n\right) = \prod_{i=1}^{N} \frac{1}{\sqrt{2\pi\sigma_i^2}} e^{-\frac{\left(y_i - Y_i^n\left(x_i, \theta_1^n, \ldots, \theta_m^n\right)\right)^2}{2\sigma_i^2}} \tag{7.3}$$

$$\mathcal{L}\left(\theta_1^n, \ldots, \theta_m^n\right) = \left(\prod_{i=1}^{N} \frac{1}{\sqrt{2\pi\sigma_i^2}}\right) e^{-\sum_{i=1}^{N} \frac{\left(y_i - Y_i^n\left(x_i, \theta_1^n, \ldots, \theta_m^n\right)\right)^2}{2\sigma_i^2}} \tag{7.4}$$

$$\mathcal{X}^2\left(\theta_1^n, \ldots, \theta_m^n\right) = \sum_{i=1}^{N} \left(\frac{y_i - Y_i^n\left(x_i, \theta_1^n, \ldots, \theta_m^n\right)}{\sigma_i}\right)^2 \tag{7.5}$$

$$\mathcal{L}\left(\theta_1^n, \ldots, \theta_m^n\right) = \left(\prod_{i=1}^{N} \frac{1}{\sqrt{2\pi\sigma_i^2}}\right) e^{-\frac{\mathcal{X}^2\left(\theta_1^n, \ldots, \theta_m^n\right)}{2}} \tag{7.6}$$

Then, a candidate set for the model parameters $\left(\left\{\widetilde{\theta_1^{n+1}}, \ldots, \widetilde{\theta_m^{n+1}}\right\}\right)$, as the next link, is randomly chosen using the proposal distribution function, and its likelihood value $\mathcal{L}\left(\widetilde{\theta_1^{n+1}}, \ldots, \widetilde{\theta_m^{n+1}}\right)$ is determined.

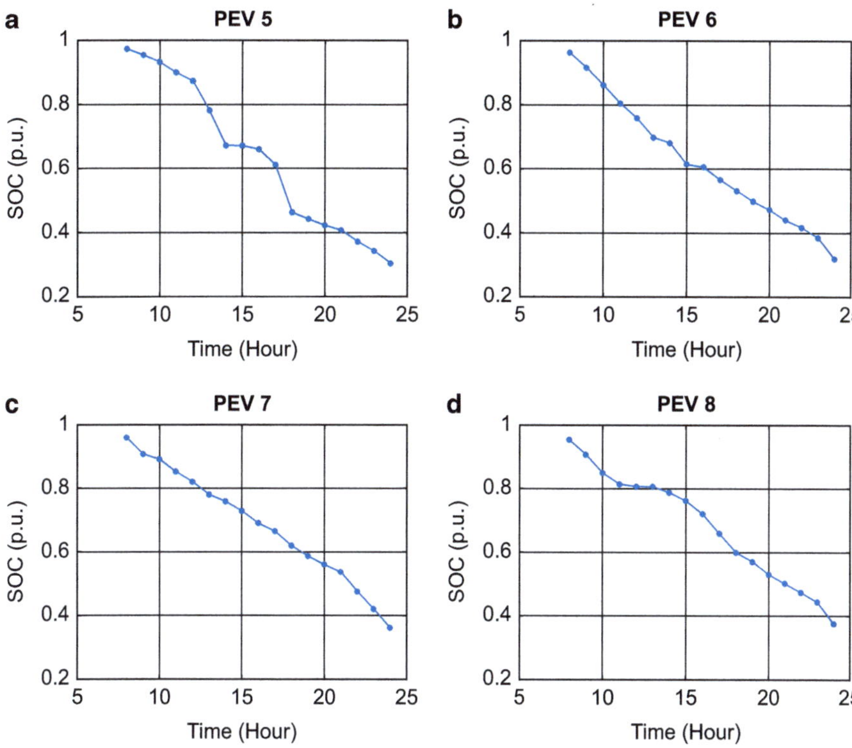

Fig. 7.16 The SOC (p.u.) of PEVs 5–8 estimated at each hour of period of 8:00–24:00

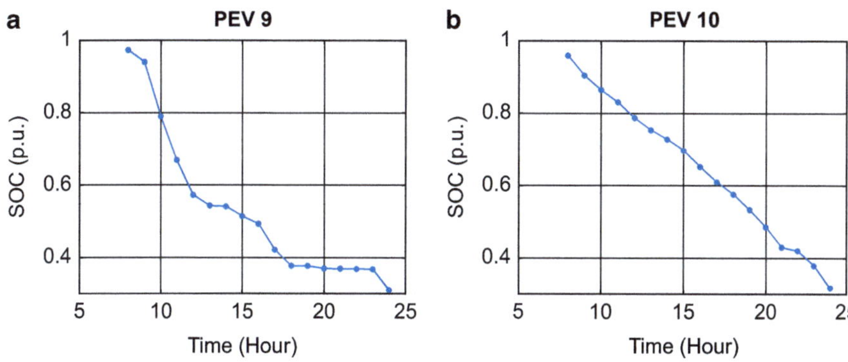

Fig. 7.17 The SOC (p.u.) of PEVs 9–10 estimated at each hour of period of 8:00–24:00

Afterward, the acceptance or rejection of candidate set of model parameters is specified using (7.7)–(7.8). As can be seen, if the likelihood value of candidate set of model parameters ($\mathcal{L}(\widetilde{\theta_1^{n+1}}, \ldots, \widetilde{\theta_m^{n+1}})$) is larger than the one of current set of model parameters ($\mathcal{L}(\theta_1^n, \ldots, \theta_m^n)$), the candidate set ($\widetilde{\theta_1^{n+1}}, \ldots, \widetilde{\theta_m^{n+1}}$) is accepted as the new

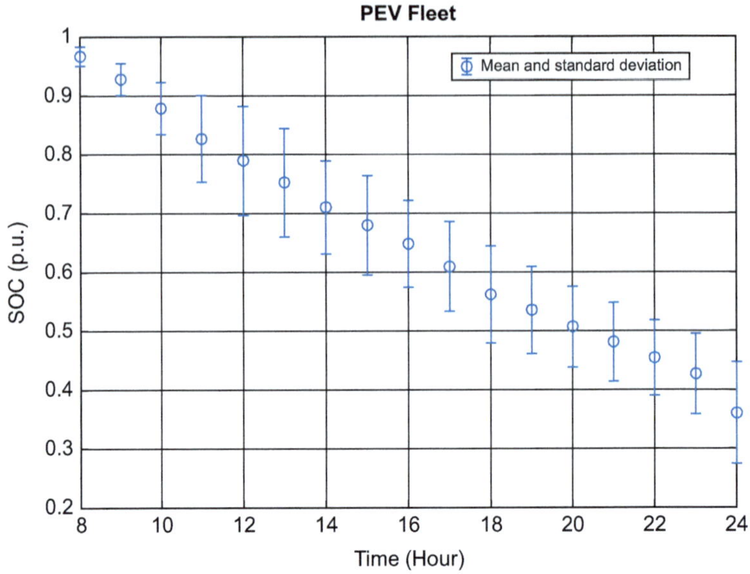

Fig. 7.18 The mean and standard deviation of SOC (p.u.) of PEV fleet (PEVs 1–10) estimated at each hour of period of 8:00–24:00

link. Moreover, the candidate set of model parameters has a chance to be accepted as the new link if the ratio of likelihood value of candidate set to the likelihood value of current set ($\alpha(\widetilde{\theta_1^{n+1}}, \ldots, \widetilde{\theta_m^{n+1}} | \theta_1^n, \ldots, \theta_m^n)$) is larger than a random number between zero and one (*rand*[0, 1)). Otherwise, the candidate set of model parameters is rejected, the MCMC does not move from its current set to the new one, and just the same set is repeated for the next link of MCMC. This procedure is repeated for a large number of iterations until the satisfactory results are achieved.

$$\begin{aligned} &\mathit{Decsion} \\ &= \begin{cases} \mathit{Acceptance} & \mathcal{L}(\widetilde{\theta_1^{n+1}}, \ldots, \widetilde{\theta_m^{n+1}}) > L(\theta_1^n, \ldots, \theta_m^n) \\ \mathit{Acceptance} & \mathcal{L}(\widetilde{\theta_1^{n+1}}, \ldots, \widetilde{\theta_m^{n+1}}) < L(\theta_1^n, \ldots, \theta_m^n),\ \alpha(\widetilde{\theta_1^{n+1}}, \ldots, \widetilde{\theta_m^{n+1}} | \theta_1^n, \ldots, \theta_m^n) > rand[0,1) \\ \mathit{Rejection} & \mathcal{L}(\widetilde{\theta_1^{n+1}}, \ldots, \widetilde{\theta_m^{n+1}}) < L(\theta_1^n, \ldots, \theta_m^n),\ \alpha(\widetilde{\theta_1^{n+1}}, \ldots, \widetilde{\theta_m^{n+1}} | \theta_1^n, \ldots, \theta_m^n) < rand[0,1) \end{cases} \end{aligned} \tag{7.7}$$

$$\begin{aligned} \alpha(\widetilde{\theta_1^{n+1}}, \ldots, \widetilde{\theta_m^{n+1}} | \theta_1^n, \ldots, \theta_m^n) &= \frac{\mathcal{L}(\widetilde{\theta_1^{n+1}}, \ldots, \widetilde{\theta_m^{n+1}})}{\mathcal{L}(\theta_1^n, \ldots, \theta_m^n)} \\ &= e^{-\left(\frac{\chi^2\left(\widetilde{\theta_1^{n+1}}, \ldots, \widetilde{\theta_m^{n+1}}\right) - \chi^2\left(\theta_1^n, \ldots, \theta_m^n\right)}{2}\right)} \end{aligned} \tag{7.8}$$

7.4 Simulation Results

7.4.1 Base Case

In this study, the total number of iterations of MCMC is considered about 20,000. In addition, since a linear model is applied in the MCMC, just two model parameters are needed ($m = 2$). The initial values of model parameters (θ_1^1 and θ_2^1) are assumed to be about 1.00 and −0.03, respectively. The maximum variation of model parameters in each iteration of MCMC is considered about 0.01.

The MCMC simulation results are presented in Figs. 7.19, 7.20, 7.21, 7.22, 7.23, and 7.24. In Figs. 7.19 and 7.20, the values of model parameters in each iteration of MCMC are illustrated. As is seen, the value of each model parameter fluctuates around a particular value and in a specific range. Table 7.2 presents the simulation outputs of MCMC in some of the iterations. These outputs include the value of model parameters, the likelihood value of current set of model parameters, the likelihood value of candidate set of model parameters, the ratio of likelihood values, and the random number generated in the range of [0,1). As can be seen, the set of model parameters is updated if the likelihood value of candidate set of model parameters is larger than the likelihood value of current set of model parameters (see iteration numbers 2, 3, 4, 103, 1003, 10003, 19997) or if the ratio of likelihood values is larger than the random number (see iteration numbers 105, 1005, 20000). Otherwise, the candidate set of model parameters is rejected, and the same set is repeated for the next link of MCMC (see iteration numbers 5, 102, 104, 1002, 1004, 10002, 10004, 10005, 19998, 19999).

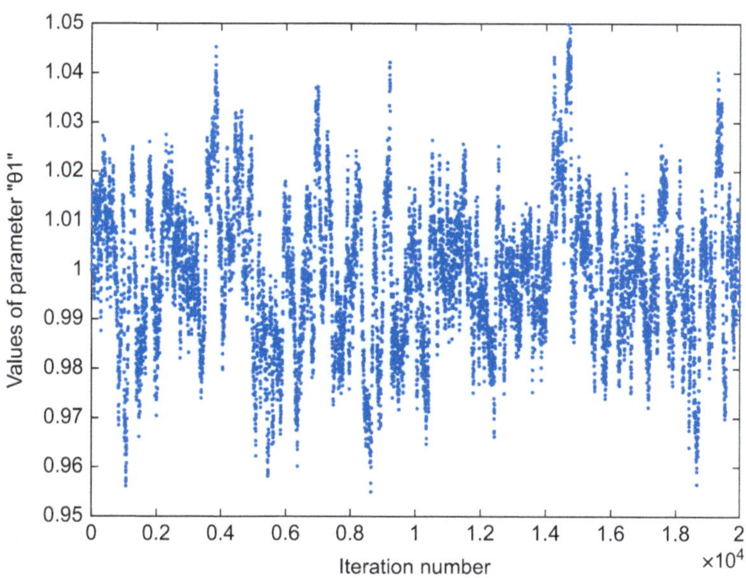

Fig. 7.19 The value of model parameter θ_1 in each iteration of MCMC

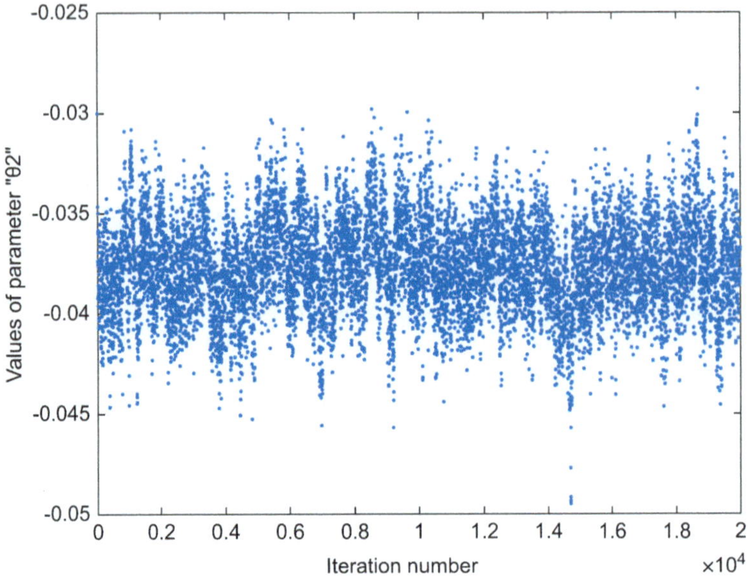

Fig. 7.20 Value of model parameter θ_2 in each iteration of MCMC

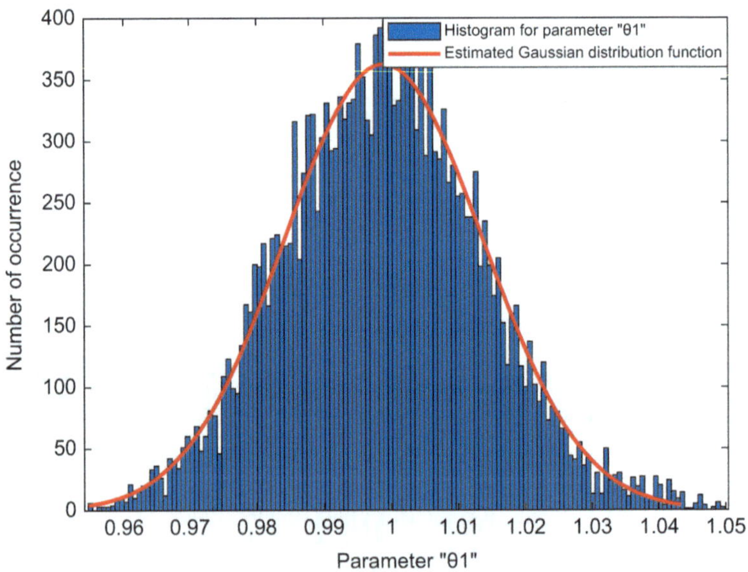

Fig. 7.21 The histogram of model parameter θ_1 and the estimated posterior distribution function

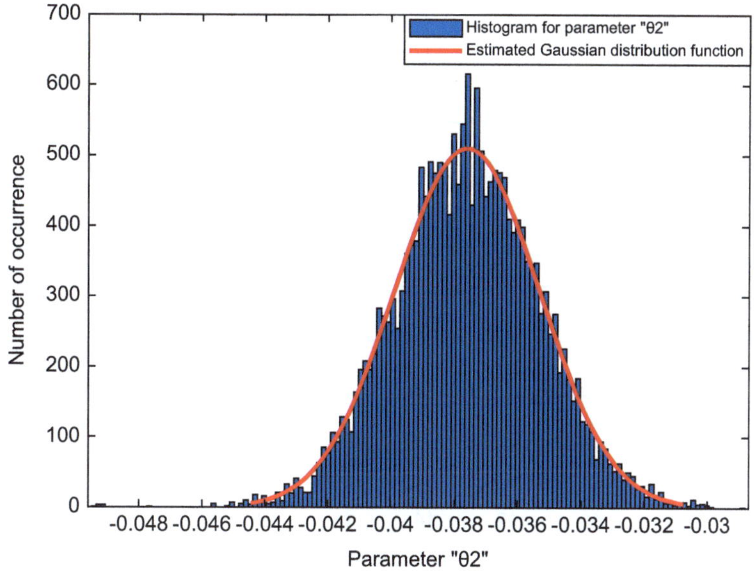

Fig. 7.22 The histogram of model parameter θ_2 and the estimated posterior distribution function

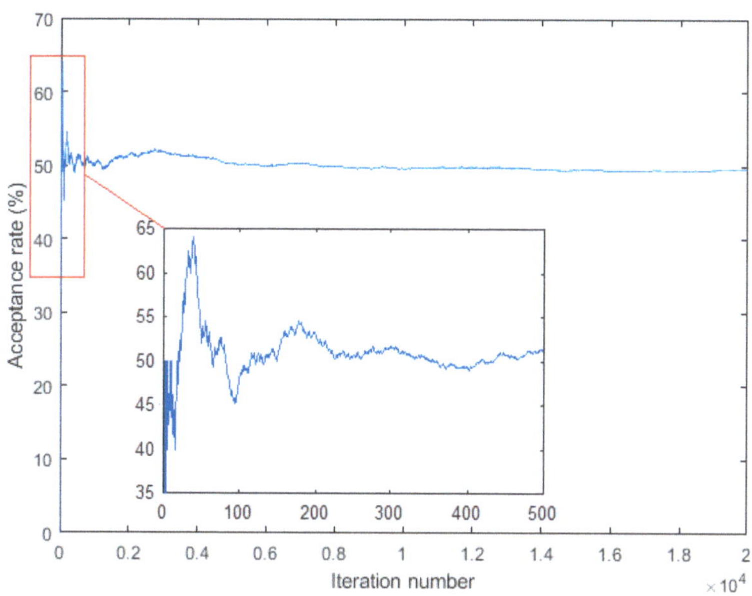

Fig. 7.23 The acceptance rate of MCMC (%) calculated up to each iteration

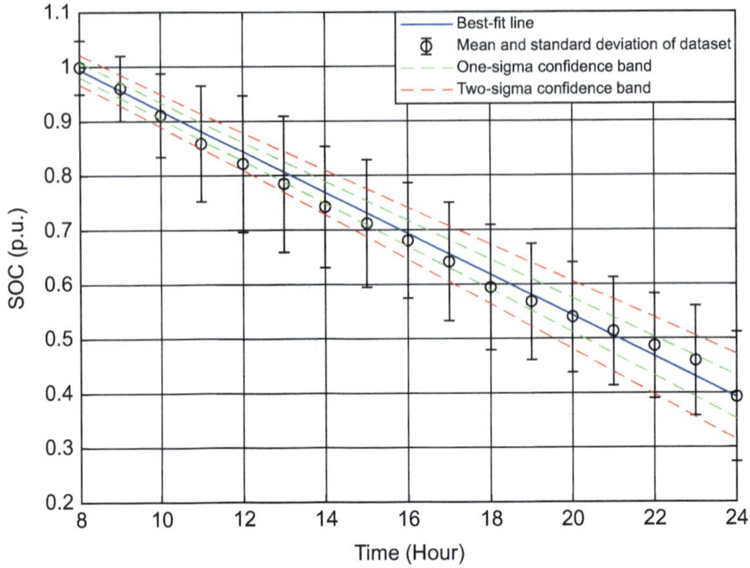

Fig. 7.24 The best-fit line, mean, standard deviation, one-sigma confidence band, and two-sigma confidence band of SOC (p.u.) of PEV fleet

In addition, Figs. 7.21 and 7.22 show the histogram of model parameters and their estimated posterior distribution functions. As can be seen, the distribution functions are approximately fitted with the Gaussian distribution functions with the mean and standard deviation of 0.998 and 0.015 for the model parameter θ_1 and -0.037 and 0.002 for the model parameter θ_2.

Moreover, the acceptance rate of MCMC (in percent), which is calculated up to each iteration, is shown in Fig. 7.23. As is observed, the acceptance rate has some sudden fluctuations for the first 200 iterations; however, it approaches 50% after that and remains almost constant until the end of MCMC simulation. The reason of this phenomenon is concerned with the initial value of model parameters.

Figure 7.24, as the desirable output of MCMC, illustrates the best-fit line, the one-sigma confidence band, and the two-sigma confidence band of SOC of PEV fleet. Herein, the mean and standard deviation of dataset have been added to Fig. 7.24. As can be seen, the best-fit line for the SOC of PEV fleet is a line that starts from 0.998 p.u. at 8th hour of the day and descends with the slope of -0.037 to reach 0.40 p.u. at 24th hour of the day. In other words, the best value of model parameters θ_1 and θ_2 are about 0.998 and $-$ 0.037, respectively.

Table 7.2 The value of model parameters, their likelihood value, ratio of their likelihood values, and random number generated in each iteration of MCMC

Iteration (n)	θ_1^n	θ_2^n	$\mathcal{L}(\theta^n) \times 10^{-11}$	$\mathcal{L}(\widetilde{\theta^{n+1}}) \times 10^{-11}$	$\alpha(\widetilde{\theta^{n+1}}\vert\theta^n)$	*rand* [0, 1)
1	1.0000	−0.0300	–	–	–	–
2	1.0014	−0.0302	0.0072	0.0089	1.2463	0.5146
3	1.0011	−0.0314	0.0089	0.1529	17.0798	0.6968
4	0.9961	−0.0335	0.1529	17.0732	111.6989	0.2989
5	0.9961	−0.0335	17.0732	0.0869	0.0051	0.9142
⋮	⋮	⋮	⋮	⋮	⋮	⋮
⋮	⋮	⋮	⋮	⋮	⋮	⋮
101	1.0033	−0.0403	86.7682	38.2168	0.4404	0.7051
102	1.0033	−0.0403	86.7682	3.7140	0.0428	0.5438
103	1.0052	−0.0392	86.7682	170.8618	1.9692	0.9423
104	1.0052	−0.0392	170.8618	9.2232	0.0540	0.2508
105	1.0047	−0.0389	170.8618	27.6897	0.1621	0.1353
⋮	⋮	⋮	⋮	⋮	⋮	⋮
⋮	⋮	⋮	⋮	⋮	⋮	⋮
1001	1.0109	−0.0374	114.5532	0.0540	0.0005	0.7877
1002	1.0109	−0.0374	114.5532	1.8934	0.0165	0.0515
1003	1.0139	−0.0398	114.5532	126.7835	1.1068	0.5625
1004	1.0139	−0.0398	126.7835	2.6627	0.0210	0.1769
1005	1.0135	−0.0396	126.7835	13.8271	0.1091	0.0905
⋮	⋮	⋮	⋮	⋮	⋮	⋮
⋮	⋮	⋮	⋮	⋮	⋮	⋮
10001	1.0249	−0.0396	49.7680	8.0990	0.1627	0.9503
10002	1.0249	−0.0396	49.7680	2.7100	0.0545	0.0587
10003	1.0207	−0.0383	49.7680	54.8337	1.1018	0.2896
10004	1.0207	−0.0383	54.8337	0.1032	0.0019	0.7764
10005	1.0207	−0.0383	54.8337	12.0585	0.2199	0.6109
⋮	⋮	⋮	⋮	⋮	⋮	⋮
⋮	⋮	⋮	⋮	⋮	⋮	⋮
19996	1.0081	−0.0372	135.5535	1.6516	0.0122	0.5670
19997	1.0106	−0.0392	135.5535	161.2034	1.1892	0.5314
19998	1.0106	−0.0392	161.2034	7.0608	0.0438	0.2779
19999	1.0106	−0.0392	161.2034	44.4283	0.2756	0.5599
20000	1.0067	−0.0369	161.2034	131.5347	0.8160	0.3008

7.4.2 Studying the Effect of Parameters

7.4.2.1 Initial Value of Model Parameters

In this part, the initial value of model parameters is changed to $\theta_1^1 = 0.8$ and $\theta_2^1 = 0$, and then the problem is simulated. As can be seen in Figs. 7.25 and 7.26, due to initiating the MCMC with the abovementioned model parameters, a burn-in process with the corresponding burn-in period appears in the MCMC of each model parameter. However, this part of MCMC is not favorable and should be removed from it. As is seen, although the initial value of model parameters affects the MCMC

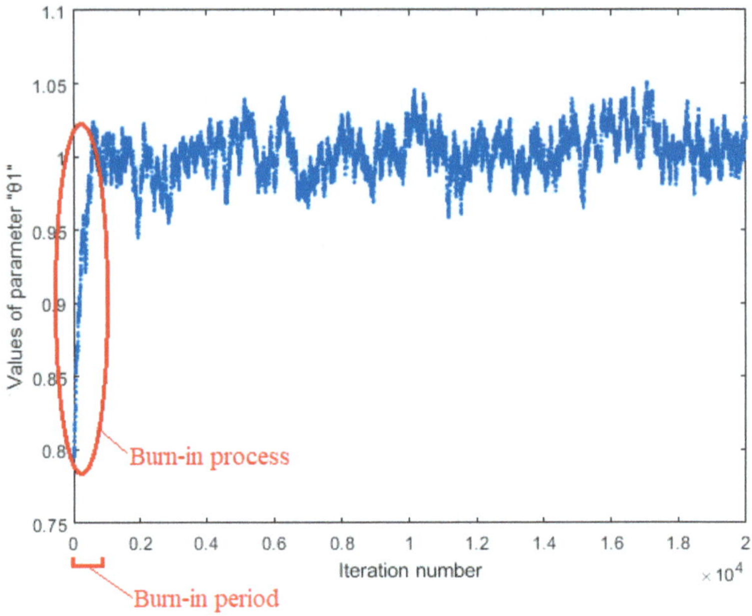

Fig. 7.25 The existence of burn-in process in the chain of θ_1 due to starting the MCMC with the initial value of $\theta_1^1 = 0.8, \theta_2^1 = 0$

procedure and causes a delay for the convergence of MCMC, the MCMC finally finds its way, and the best value of model parameters is not biased by the starting point of chain.

Moreover, Figs. 7.27 and 7.28 illustrate the effect of initial value of model parameters ($\theta_1^1 = 0.8,\ \theta_2^1 = 0$) and burn-in process on the histogram of model parameters. As is noticed, the burn-in process exacerbates the histogram of model parameters and negatively affects the shape of estimated Gaussian distribution functions.

7.4.2.2 Maximum Variation of Model Parameters

In this section, the maximum variation of model parameters in the MCMC is increased to 0.05. As can be seen in Fig. 7.29, the graphs of value of model parameters versus the iteration number become sparse, and the acceptance rate of MCMC decreases to about 10%. In other words, when the model parameters change with the larger value, the precision of MCMC decreases, the acceptance rate reduces, and the previous links are repeated.

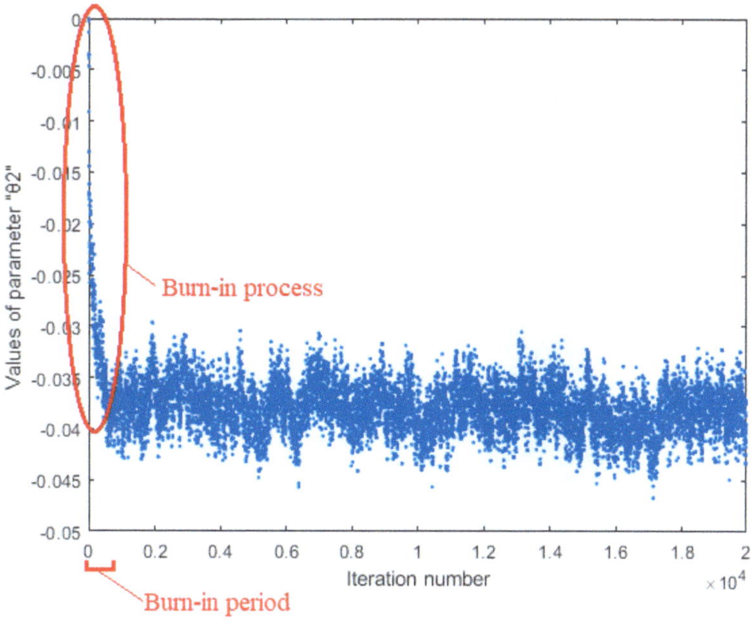

Fig. 7.26 The existence of burn-in process in the chain of θ_2 due to starting the MCMC with the initial value of $\theta_1^1 = 0.8$, $\theta_2^1 = 0$

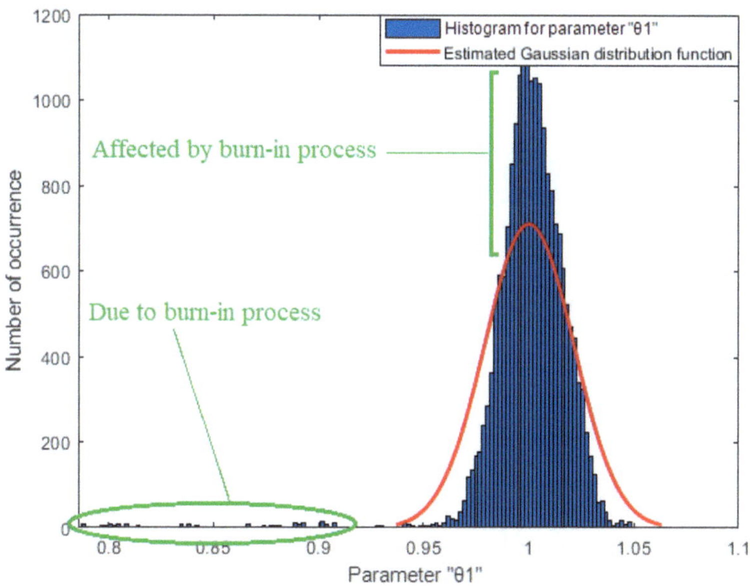

Fig. 7.27 The effect of initial value of model parameters ($\theta_1^1 = 0.8, \theta_2^1 = 0$) and burn-in process on the histogram of model parameter θ_1

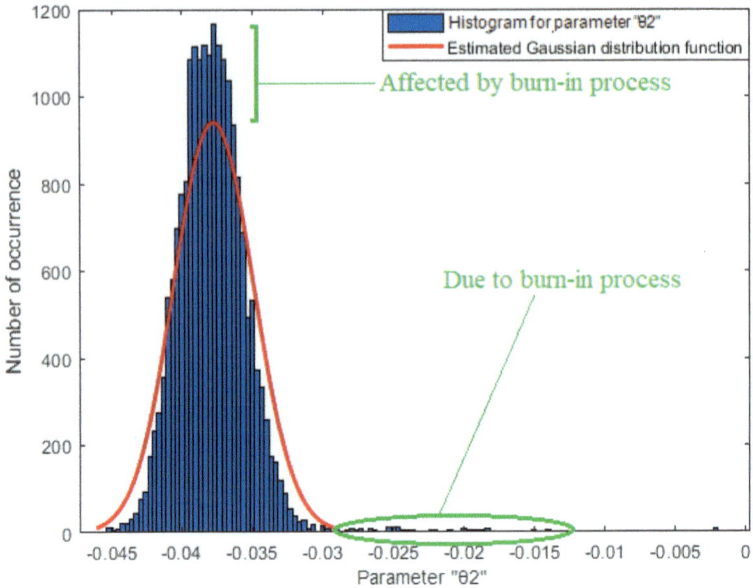

Fig. 7.28 The effect of initial value of model parameters ($\theta_1^1 = 0.8, \theta_2^1 = 0$) and burn-in process on the histogram of model parameter θ_2

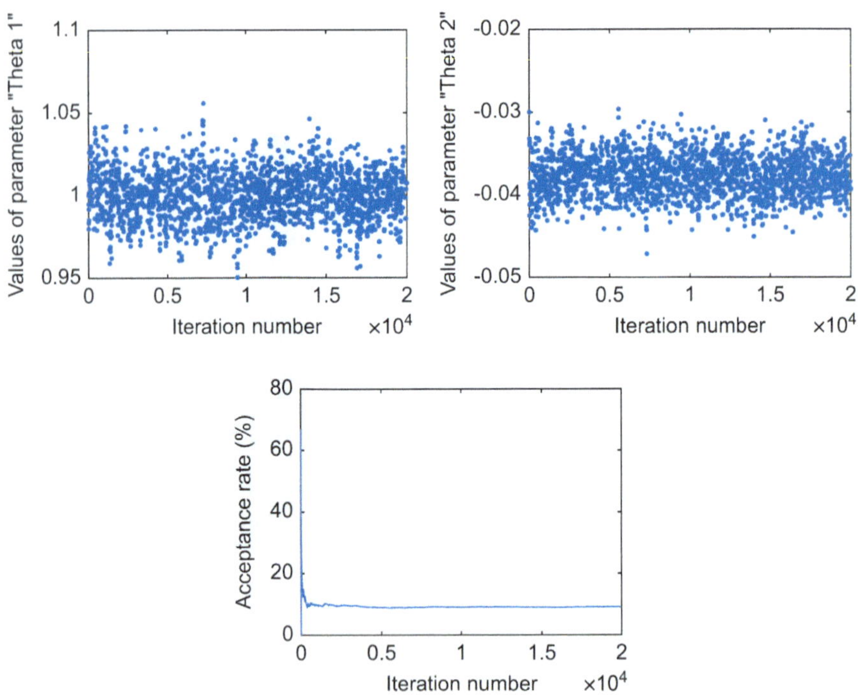

Fig. 7.29 The simulation results for the larger maximum variation of model parameters

7.4.2.3 Total Number of Iterations

In this part, the MCMC is applied for the smaller and larger number of iterations. Figures 7.30 and 7.31 show the simulation results for 2000 and 200000 iterations of MCMC, respectively. As can be seen in Fig. 7.30, the MCMC does not have enough iterations and chains to build the histogram and Gaussian distribution function of model parameters. However, the MCMC with the larger number of iterations generates a precise histogram and Gaussian distribution function for each model parameter, as is seen in Fig. 7.31.

7.4.2.4 PEVs with Different Technical Specifications

In this part, the problem is simulated considering the other PEV types that include Nissan Leaf S, Mia Electric C3, and Tesla Model S. As can be seen in Table 7.1, each model of PEV has a different energy consumption index and battery capacity. Figure 7.32 illustrates the hourly best-fit line for the SOC of PEV fleet for each PEV

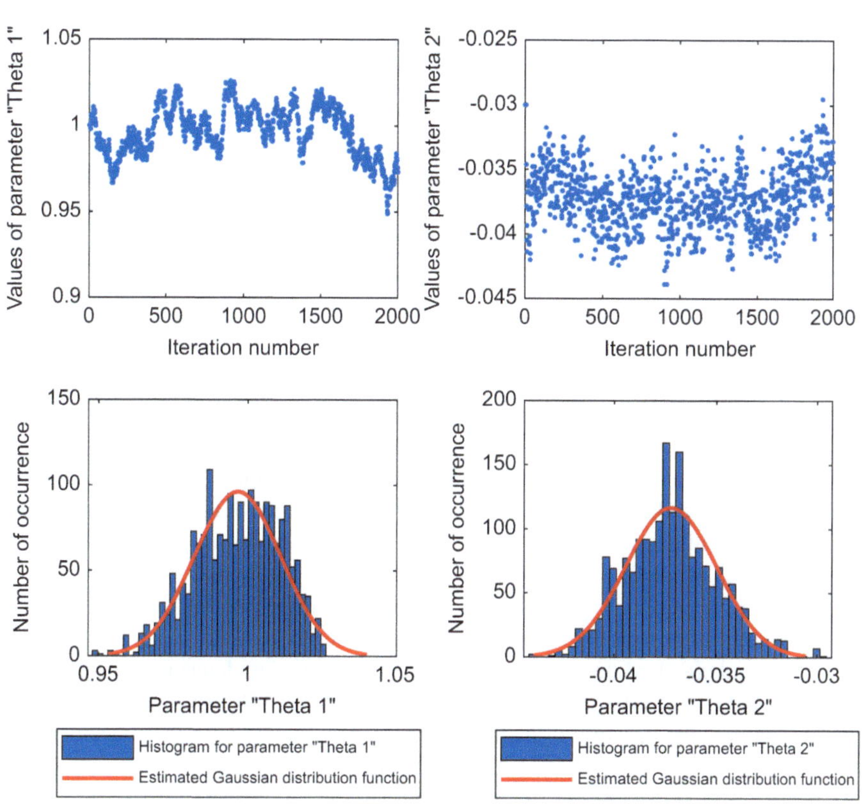

Fig. 7.30 The simulation results for the smaller number of iterations

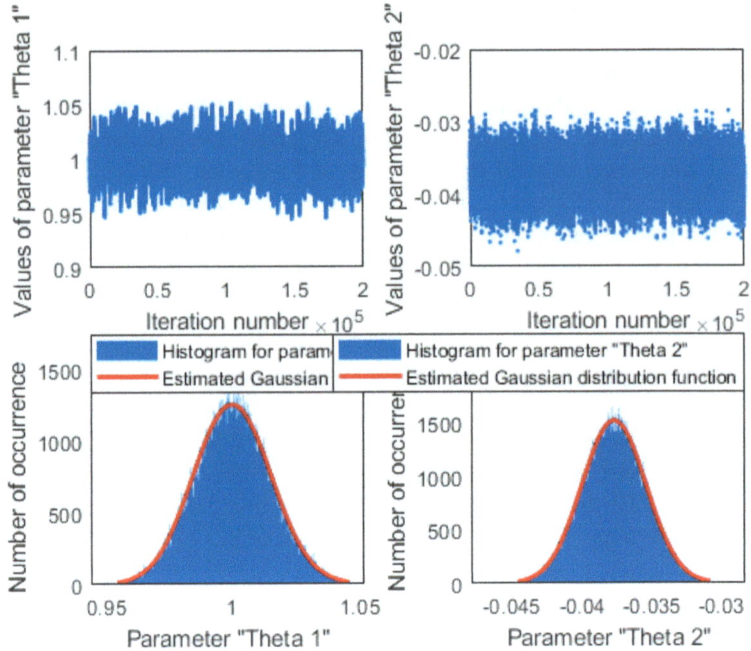

Fig. 7.31 The simulation results for the larger number of iterations

model. As can be seen, the type of PEV affects the best value of model parameter θ_2 (slope) and consequently the corresponding best-fit line. As can be noticed from Fig. 7.32, the hourly SOC of PEV fleet has an inverse relation with the energy consumption index of PEV and a direct relation with its battery capacity.

7.5 Conclusion

In this study, Monte Carlo Markov Chain (MCMC) was applied to estimate the best-fit line and the confidence bands of hourly state of charge (SOC) of plug-in electric vehicles (PEV) fleet in the day. The primary dataset used in this study included the driving routes of ten PEVs in San Francisco indicated by the real longitude (degree) and latitude (degree) of PEVs recorded in every 4-minute interval during the period of 8:00–24:00. The hourly distances travelled by the PEVs were calculated, and then the mean and standard deviation of SOC of PEV fleet were determined considering the technical characteristics of PEVs including the initial SOC, the energy consumption index, and the capacity of battery of PEVs. The time in hour was considered as the independent variable of dataset, and the mean and standard deviation of SOC of PEV fleet were considered as the dependent variable of dataset of MCMC.

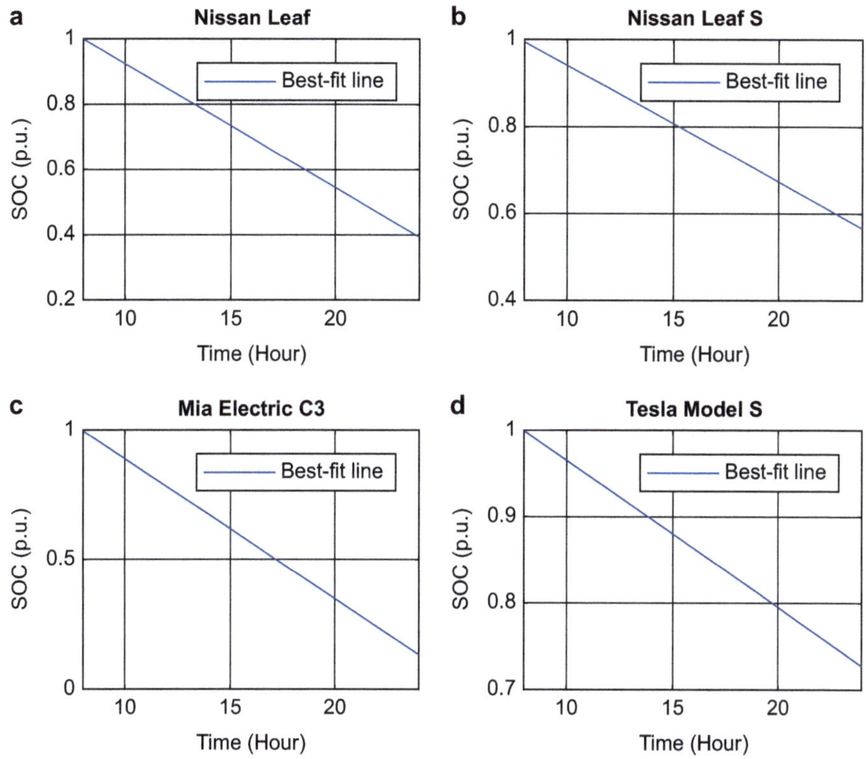

Fig. 7.32 The simulation results for different PEV types

After applying the MCMC, the posterior distribution functions of model parameters were determined, and it was noticed that the posterior distribution functions are fitted with the Gaussian distribution functions with the mean and standard deviation of 0.998 and 0.015 for the model parameter θ_1 and -0.037 and 0.002 for the model parameter θ_2. Using these results, the best-fit line, the mean and standard deviation of SOC of PEV fleet, and the one-sigma and two-sigma confidence bands were determined.

It was demonstrated that the initial value of model parameters, as the starting point of chain, can affect the MCMC process and cause a delay for its convergence during the burn-in process; however, the MCMC is able to discover the best value of model parameters and the corresponding best-fit line. In other words, the MCMC is not biased by the starting point of chain if it includes enough iterations. It was suggested to remove the links generated during the burn-in process, since they had undesirable effects on the histogram of model parameters and the estimated Gaussian distribution functions.

It was proven that, to extract the desirable results from the MCMC in an acceptable period, the parameters of MCMC including the maximum variation of model parameters and the total number of iterations need to be appropriately adjusted. Otherwise, the precision of MCMC simulation results will be negatively affected, and the MCMC will not have enough iterations and chains to construct the probability distribution function of model parameters.

It was shown that the type of PEV can influence the best-fit line of SOC of PEV fleet, where the best-fit line has inverse and direct relations with the energy consumption index and battery capacity of PEV, respectively.

Nomenclature

α	Ratio of likelihood values in MCMC
θ^1	Initial value of set of model parameters in MCMC
θ^n	Current set of model parameters in MCMC
$\widetilde{\theta^{n+1}}$	Candidate set of model parameters in MCMC
σ^2	Variance of dependent variable of dataset in MCMC
χ^2	Chi-squared variable defined in MCMC
C^{PEV}	Capacity of battery of a PEV (kWh)
$Decsion$	Decision to accept or reject the new link in MCMC
$Distance^{PEV}$	Distance travelled by a PEV
kWh_{km}	Energy consumption index of a PEV (kWh/km) which is the amount of energy (kWh) that a PEV consumes to travel about 1 km
$\mathcal{L}$	Likelihood value in MCMC
i, N	Index of data and size of dataset (total number of sample data) in MCMC, respectively
$rand[0, 1)$	A random number between zero and one used in MCMC
SOC_0^{PEV}	Initial state of charge of a PEV before starting its commute
SOC^{PEV}	State of charge of a PEV
t, τ	Indices of time
x, y	Independent variable of dataset and dependent variable of dataset in MCMC, respectively
Y^n	Current value of dependent variable of model in MCMC
x^{PEV}, y^{PEV}	Position data of a PEV

References

1. [Online]. Available: https://www.ucsusa.org/clean-vehicles/electric-vehicles#.W_wbXehKjIU. Accessed on Nov 2018
2. [Online]. Available: https://insideevs.com/monthly-plug-in-sales-scorecard/. Accessed on Jan 2019
3. [Online]. Available: https://www.bloomberg.com/graphics/2017-lithium-battery-future/. Accessed on Jan 2019
4. M. Rahmani-Andebili, M. Fotuhi Firuzabad, An adaptive approach for PEVs charging management and reconfiguration of electrical distribution system penetrated by renewables. IEEE Trans. Ind. Inf. **14**(5), 2001 (2018)

5. M. Rahmani-Andebili, H. Shen, M. Fotuhi Firuzabad, Planning and operation of parking lots considering system, traffic, and drivers behavioral model. IEEE Trans. Syst. Man Cybern. Syst. (2018). https://doi.org/10.1109/TSMC.2018.2824122
6. M. Rahmani-Andebili, H. Shen, Traffic and grid-based parking lot allocation for PEVs considering driver behavioral model, in *International Conference on Computing, Networking and Communications (ICNC)*, Silicon Valley, USA, 26–29 Jan 2017
7. [Online]. Available: https://crawdad.org/all-byname.html. Accessed on Nov 2018
8. [Online]. Available: https://www.usgs.gov/faqs/how-much-distance-does-a-degree-minute-and-second-cover-your-maps. Accessed on Nov 2018
9. D. Newbery, "The Economics of Electric Vehicles," EPRG and Imperial College London, E&E Seminar, Cambridge, Jan 2013. Accessed on Nov 2018. [Online]. Available: http://www.eprg.group.cam.ac.uk/wp-content/uploads/2013/01/EEJan13_EconomicsEVs.pdf
10. [Online]. Available: http://www.elektriauto.ee/eng. Accessed on Nov 2018
11. M. Bonamente, Chapter 16: Monte Carlo Markov chains, in *Statistics and Analysis of Scientific Data*, 2nd edn., (Springer New York Heidelberg Dordrecht London, Springer, 2017). https://doi.org/10.1007/978-1-4939-6572-4
12. S.M. Ross, *Introduction to Probability Models* (Academic, San Diego, 2003)
13. W.K. Hastings, Monte Carlo sampling methods using Markov chains and their applications. Biometrika **57**(1), 97–109 (1970). https://doi.org/10.1093/biomet/57.1.97. http://biomet.oxfordjournals.org/content/57/1/97.abstract
14. N. Metropolis, A.W. Rosenbluth, M.N. Rosenbluth, A.H. Teller, E. Teller, Equation of state calculations by fast computing machines. J. Chem. Phys. **21**, 1087–1092 (1953). https://doi.org/10.1063/1.1699114
15. N. Gehrels, Confidence limits for small numbers of events in astrophysical data. Astrophys. J. **303**, 336–346 (1986). https://doi.org/10.1086/164079
16. G. Cowan, *Statistical Data Analysis* (Oxford University Press, Oxford, 1998)
17. W. Press, S. Teukolski, W. Vetterling, B. Flannery, *Numerical Recipes 3rd Edition: The Art of Scientific Computing* (Cambridge University Press, Cambridge, 2007)

Index

M. Rahmani-Andebili, *Planning and Operation of Plug-In Electric Vehicles*,
https://doi.org/10.1007/978-3-030-18022-5

E

MIX
Papier aus verantwortungsvollen Quellen
Paper from responsible sources
FSC® C105338

If you have any concerns about our products,
you can contact us on
ProductSafety@springernature.com

In case Publisher is established outside the EU,
the EU authorized representative is:
Springer Nature Customer Service Center GmbH
Europaplatz 3, 69115 Heidelberg, Germany

Printed by Libri Plureos GmbH
in Hamburg, Germany